THREE DAUGHTERS
THREE JOURNEYS

THREE DAUGHTERS
THREE JOURNEYS
Quest for Cancer Cure

Ananda M. Chakrabarty
Jill Charles
Indrani Mondal
Ranjita Chattopadhyay

PAN STANFORD PUBLISHING

Published by

Pan Stanford Publishing Pte. Ltd.
Penthouse Level, Suntec Tower 3
8 Temasek Boulevard
Singapore 038988

Email: editorial@panstanford.com
Web: www.panstanford.com

British Library Cataloguing-in-Publication Data
A catalogue record for this book is available from the British Library.

Three Daughters, Three Journeys: Quest for Cancer Cure

ISBN 978-981-4745-90-1 (Hardcover)
ISBN 978-1-315-19666-4 (eBook)

Printed in the USA

Contents

Preface

Cancer is a dreaded disease worldwide with countless victims. Although cancers in different organs may have different etiologies, essentially all cancers are due to mutations in our genome, anywhere in a few to 100 different genes. The mutations could be due to lifestyle choice or simply due to bad luck. Several such mutations have been found in two genes known as BRCA1 and BRCA2 genes, occurring in two different chromosomes in our bodies, which may lead to the development of breast or ovarian cancer in women at a much higher frequency than normal. Inheritance of such BRCA1 and/or BRCA2 gene mutations from the parents can therefore lead to breast or ovarian cancer in women with family history of such cancers at a much higher rate, thereby causing alarm in young women with such a family history. This book is a fiction based on three young girls with family histories of breast or ovarian cancer who were diagnosed to harbor the harmful mutations. The three young girls, living on three different continents, however, pursued three different avenues for their therapy, which is the focus of this book. Although the paths followed by the second and third daughters are fictional at this time, the book discusses the future potential of the development of such therapies for cancer, as reflected during the conversations of the physicians, to provide some glimpse of how the future therapy of cancer might take shape. The authors would love to hear our readers' thoughts and concerns about the message, fictional as it is, implied in this book. The stories of the three daughters have been penned by Jill Charles ("Story of Selena"), Indrani Mondal ("Story of Marzena") and Ranjita Chattopadhyay ("Story of Kamola"), with Jill Charles writing up the concluding sections (Chapters 19 and 20) and editing the book.

Ananda M. Chakrabarty

1

SELENA

Sunflowers

SELENA ALWAYS REMEMBERED HER mother's sunflowers. Nine feet tall, bowing their golden heads toward the lawn, with yellow goldfinches hiding in their green leaves, nipping their striped seeds. In her earliest memory, Selena toddled over to the giant plants, tried to climb them, and cried when the thick fuzzy stems bent and broke under her weight. These flowers stood taller than she was, taller than Mama, even taller than Daddy, a long lean man people often mistook for a basketball player.

When Selena couldn't climb the sunflowers, her mama and daddy lifted her up on their shoulders so she could stroke the soft yellow petals. Later they taught her how to crack the seeds in her teeth and spit the shells.

Three Daughters, Three Journeys: Quest for Cancer Cure
Ananda M. Chakrabarty, Jill Charles, Indrani Mondal,
and Ranjita Chattopadhyay
Copyright © 2017 Pan Stanford Publishing Pte. Ltd.
ISBN 978-981-4745-90-1 (Hardcover), 978-1-315-19666-4 (eBook)
www.panstanford.com

The brown and gold of the sunflowers reminded Selena of her mama's soft brown skin and the gold she liked to wear in earrings, necklaces and tiny rings at the ends of her long black braids. The height of the flowers and the way they swayed in the Chicago wind reminded her of Daddy playing blues guitar, at home or at his club, the Blue Jay.

The Blue Jay had three floors with DJs, a 24-hour restaurant, two bars, a dance club and a concert hall. Selena would run around after Daddy, tapping the cymbals or dancing to the piano while the bands tuned up. She could look out the club windows and see Lake Michigan, Chicago's skyline with the Sears Tower, Hancock Building and the El train tracks of the South Loop. Daddy pointed out Union Station, where he had arrived in town years ago on a train from New Orleans, with only a steel guitar, harmonica, and backpack full of twenty-year-old dreams. He'd played in eight bands—blues, jazz and rock—and worked his way up to managing a club, managing an apartment, then buying a South side bar, then an apartment complex. He and Mama bought an old Bronzeville mansion and remodeled it, like a gray stone palace.

Daddy called Selena "Princess" and her mother "Queen Bea." Unlike her husband Jack Ramos, Beatrice Wells Ramos came from a well-to-do black family that had lived in Bronzeville since the 1940s. She was the third generation in her family to go to college, in a long line of preachers, teachers, businessmen and doctors. Beatrice was majoring in history at the University of Chicago when she met Jack. She saw him playing *Walking Blues* in a dark, smoky club and fell completely in love with him. He looked out into the audience and spotted her, with her gold-tipped braids and a black-eyed Susan flower behind her ear, and he fell in love with Bea too. They danced together before they spoke, then talked all night, just the beginning of a conversation that would last a lifetime. They married right after Beatrice graduated and Selena came along two years later.

Selena always wished for a sibling, but her busy parents struggled to make time for one daughter as her mother taught women's studies and Black history and competed hard for tenure. Her dad invested more in real estate and hired new musical acts and DJs at the Blue Jay every month. Selena felt lucky when she saw how her parents stretched their schedules, to make sure at least one of them shared breakfast with her, cooking her favorite soft-boiled egg and oatmeal. At least one of them would pick her up from school every day and help her with long division and Spanish verbs. Daddy would get up with Selena after being up till 3 a.m. playing blues when a musician called in sick to the club. Mama would pile her history books, some of which she wrote, on the kitchen table with Selena's. They drank mint tea while they did homework together.

To keep Selena company, her parents always let her have pets: two cockatiels named Billie and Buddy, a fluffy black chinchilla named Coco and a huge aquarium of angelfish, mollies and tetras, each of whom Selena had named and fed and cleaned up after every day. Selena tried to take care of her pets like her parents took care of her. At age ten she was allowed to adopt a dog and chose a rescued black Greyhound, free to a good home, who she named Jojo. Jojo loved Selena so much she'd wait inside the garden gate for her to come home from school even when snow covered their garden. Selena loved to take the Green Line El train home from her selective enrollment public high school and see Jojo waiting to play in the yard and walk around their neighborhood.

Selena made friends eagerly with all kinds of children: girls, boys, outspoken class clowns, shy bookish kids who rarely spoke except to her, basketball stars, cheerleaders and a deaf girl who made rap videos in sign language. Since kindergarten, Selena had kept the same best friend, Janine Jones, a plump girl with skin the color of chocolate milk and big glasses who loved to draw cartoons and even made a comic book version of their lives. Selena wanted to become blood sisters, but Janine dreaded needles and would have none of it.

"We're sisters in our souls. That's all we need," Janine said.

Life was laughter, love and adventure, just like Janine's comics, until Selena's seventeenth Christmas. Mama was in the kitchen putting the final glaze on a ham crowned with pineapple rings and cloves. Aromas of collard greens with smoked turkey, cornbread, sweet potatoes and macaroni and cheese filled the house. Two of Mama's pies, sweet potato and pecan, cooled on the counter. Daddy was picking up his mother, Grandma Grace, at the airport. Their friends and the aunts, uncles and cousins would all be over soon. Janine had arrived first and was helping Selena set the table with golden napkins and red glass plates.

From the dining room, Selena heard her mother gasp and something clatter to the kitchen floor. She rushed in to help Mama.

"Mama, are you OK?"

Her mother leaned over the kitchen sink, clutching the counter. She had dropped a wooden spoon and Selena bent down to pick it up. Mama pressed a hand, hard, into the left side of her belly, between her hipbone and her navel.

She whispered, "I'm all right, baby. Just some cramps."

"Sit down for a minute," Selena said. "I'll get the food ready."

"No, no," Mama insisted. "I'm all right."

Janine and Selena carried all the heavy serving dishes, wanting to help, but unable to convince Mama to rest. Mama looked more tired than usual, even when she hugged Grandma Grace and kissed Daddy, as she welcomed all their friends and family to dinner. She tried to hide her pain but Selena felt it.

A few nights after Christmas, Selena woke up to get a drink of water. As she crossed the upstairs hall, she heard her parents talking downstairs.

Daddy hugged Mama in the rainbow glow of their Christmas tree lights.

"You take care of everybody, Queen Bea. It's my job to take care of you this time."

"Jack, you know how I hate doctors."

"I know. That's why I made you an appointment with Dr. Chandra Patel. She's Dahlia Jones's doctor and the absolute best, plus being a U of Chicago graduate."

"I think I had her in freshman history. Jack, I don't want to see someone younger than I, someone who remembers me as a professor, not even if she's your star performer's favorite doctor."

"Then you can change doctors, Bea, but the appointment's on Thursday and I'm going with you."

"What about Selena?"

"Mom's taking her shopping downtown all day. We won't need to worry her with any of this until we get the test results. It could be nothing, but you have to be checked."

"I'll see Dr. Patel."

Too scared to confront her parents, Selena lay down in her bed and cried softly. Mama and Daddy seldom disagreed strongly on anything. She could tell her mother was scared, not angry, and she already suspected why.

Selena turned eighteen in January. A week afterward, she sat in the living room drinking tea with her mother. She had convinced herself that whatever tests her mother had must have come back negative and been forgotten. Then, instead of asking her what her homework was, Mama said, "Have tea and talk with me."

Sensing that this would be a serious talk, she settled beside her mother on the couch.

"Selena, you know I've been seeing Dr. Patel."

"Yes."

"I've been having abdominal pain for a few months and your dad thought I should get it checked out."

"Right."

Selena set her teacup down. Her voice sounded small and her face felt hot.

"Dr. Patel wanted to be certain about my symptoms and the cause. I had an ultrasound, an MRI and a biopsy."

Selena held Mama's hand, something she hadn't done since grade school. Her mother looked down, then looked Selena in the eye.

"Selena, I have cancer. There are two tumors, in my left ovary and in my left breast. It's stage III and it has metastasized."

"What?"

"That means the cancer is in my lymph nodes and has spread to more than one part of my body. I'll have a double mastectomy and have both my ovaries removed, then have chemotherapy and radiation. Treatment will be hard, but I know I can count on Daddy and you to help me."

"Of course, but why do you have to have both breasts and both ovaries taken out if only one of each is cancerous?"

Mama let go of Selena's hand and sipped her tea.

"This type of cancer has a genetic element. It's caused by two genetic mutations on the BRCA1 and BRCA2 genes. I tested positive for both mutations. Women who have one or both are very likely to get breast or ovarian cancer. I got a second opinion from another doctor, and she agreed it's safest for me to have both breasts and both ovaries removed to minimize the risk of more tumors."

"Mama, does anyone else in our family have this?"

"No one else in our family has been tested, but you know we lost Grandma Lily to breast cancer when you were four."

Selena nodded. She had a few happy memories of Mama's mother holding her, singing and reading to her.

"If it's genetic, will I have it?"

Mama covered her face with her hands and burst into tears. Selena hugged her, scared and miserable. She had never seen her mother cry.

"Mama, I'm so sorry. What can I do to help?"

Both of them cried softly now.

"Stick with your schoolwork. Stick with your college applications."

"I'm not going away to Spellman or Stanford. I'll go to the University of Chicago."

"I don't want you missing out on anything for me. I especially don't want you to drink or smoke or try drugs to kill the pain. This will be bad, but it won't be forever. In one year, my chemo will be done."

"I'm not leaving town, Mama. I can ace all my classes, but I'm not leaving you for one minute. I can help drive you to your appointments. I can cook and clean house."

"Shh. I don't want you missing school for me. Though I have to say I'm thrilled that you offered to clean up. You always help Daddy and me, and Grandma Grace will come to live with us for a while. We'll get through this, Selena. Tomorrow we can go out and buy me a wig, maybe a purple one."

Selena tried to laugh.

It hurt to see her mother walking slowly, hurt to find her asleep, exhausted on the couch after chemo. Sometimes Selena could barely wake Mama up with shaking and shouting, which terrified her. It shocked her to see Mama take off her wig (natural black braids tipped with gold, not punk purple) and scratch her bare scalp, then wrap a cotton scarf around her head.

Mama rarely complained about anything, though Daddy, Selena and Grandma Grace saw how she suffered. They rushed to help her so much that Mama protested, "Stop fussing over me."

"Daddy, I'm so scared for her," Selena confided.

Daddy hugged her and said, "Me too, Princess. But she's a fighter. And Dr. Patel is the best oncologist in town, one of the best in the country. We'll get through this. I'm so proud of you looking after Mama. You're a real adult now."

Adulthood meant more than the independence Selena always hoped for; it meant loss and pain and never, ever taking love for granted again.

Having Grandma Grace move into their biggest guest room was the only positive aspect of Mama's cancer. Grandma Grace helped with everything at home. She drove Mama to and from appointments or stayed with Selena while Daddy took Mama to chemo. Daddy would have to carry her upstairs and tuck her into bed, like a baby.

Selena prayed harder than she'd ever prayed before. She promised God she would be perfect in mind, word and deed if only Mama could survive. She remembered all the times she'd been rude to Mama or ignored her or wanted to go out with Janine and her friends instead of staying home with her family. Now Selena felt she could never have enough time with her mother. Had she ever thanked God for giving her such a good mother who loved her and took care of her every day? Although Selena always loved her mother, she realized she had never thanked God for her. Like children often do, she took her parents' love for granted and never even tried to imagine her life without it.

"Thank you for giving me a good mother," Selena prayed. "I will help her no matter what, Lord. I will make her proud of me. I'll be brave and wise and kind, like she is and use everything I learn to help others too. Please make me like Mama. Please let me keep my Mama."

Mama came to Selena's high school graduation, though she leaned heavily on Daddy and had to walk with a cane. That summer she went to stay at the hospital. The cancer had spread and entered stage IV with tumors in her liver and her right lung. Dr. Patel estimated that Mama had six months. She was too sick to go home, though Daddy could have afforded a full-time nurse and Grandma Grace was more than willing to be a full-time caregiver. Mama chose not to continue chemo and to move into a hospice, to receive palliative care and visits from all her family and friends, from many of her students. Daddy played his guitar softly in her room and sang the blues to her. Grandma Grace and Selena visited her every day with all the news of school, the university, church and their neighbors. Selena read Mama her college term papers. The painkillers clouded Mama's energy, but she was still there.

In October, Selena visited Mama in the hospital, showing her pictures of the University of Chicago dorm room she and Janine had decorated with plants and posters.

"You're doing so well," Mama whispered. "I'm so proud of you. I love you, honey."

Selena set her hand on her mother's, but couldn't hold it because of the IV.

"Mama, please come home."

"Selena, you know I can't. I don't want any room in our house to be the room where I passed, and for you and Daddy and Grandma to have to walk past it every day. I want all your memories of home to be good memories."

"Every memory with you is a good memory. Mama, please don't leave me."

"I'll never leave you, Selena. I am with you. I am in you, no matter what you do or where you go. Me and Daddy and your grandparents, we are part of you and our love will always be there. Never doubt God's love and never doubt mine. Promise me, you'll always remember that."

"I promise. Mama, why did this happen?"

"I don't know why this happens. No one deserves this, no one and we just have to love each other and help each other. It's OK to cry and to doubt and to be angry, but know you are never alone, Selena. Remember."

Beside Mama's bed, the setting sun shone on a green glass vase of sunflowers from her garden.

Selena's Mama died in November, just before Thanksgiving.

2

MARZENA

Daybreak

MARZENA INCREASED HER SPEED on the treadmill. She felt her hamstrings and calf muscles tingle and knew she'd be sore after her fast run so early in the morning. But she couldn't stop. As she ran, she looked out through the huge glass windows all around the gym. She was on the seventieth and topmost floor of a five-star hotel in downtown Seattle. The view was surreal.

Dawn was just breaking outside over the quiet waters of Elliott Bay. In the far horizon stealthy fingers of light touched the snowcapped peak of Mount Rainier, making it sparkle. In the misty morning glow Marzena remembered Gdansk, a coastal town on a gulf of the Baltic Sea in Northern Poland where she'd been born

Three Daughters, Three Journeys: Quest for Cancer Cure
Ananda M. Chakrabarty, Jill Charles, Indrani Mondal,
and Ranjita Chattopadhyay
Copyright © 2017 Pan Stanford Publishing Pte. Ltd.
ISBN 978-981-4745-90-1 (Hardcover), 978-1-315-19666-4 (eBook)
www.panstanford.com

and had spent the first seventeen years of her life. After high school, she'd won a scholarship for further studies in the USA and had been admitted to a prestigious nursing school in Upstate New York. Four years of intense theory and practical courses followed and she had been awarded the much anticipated nursing degree for R.N. last summer. Central Hospital of Upstate New York, where she had been interning, had hired her right away, to work with the pediatrician in charge, Dr. Kim Wong, as a nurse in the neonatology department. She was here in Seattle to attend a five-day neonatology nursing conference with Dr. Wong's team. She would be presenting a paper on "Acute Needle Phobia in Young Children and Remedies."

On the skyline visible from her front window, Seattle's famous Space Needle towered menacingly, or so it seemed to Marzena. On the second night of the conference, she'd gone there on a sightseeing tour with her convention team. Now she cringed inwardly as she remembered that visit. Looking up at the pointed construction, without any previous indication, she had started trembling uncontrollably, broken out in cold sweat and had almost passed out. Vivid memories of her own acute needle phobia had come racing back to her. Piercing cries of her older brother taking vaccination shots, which she'd heard as a child in Gdansk, rang relentlessly in her head. Her brother had succumbed to polio in spite of being forced to take those painful injections and the irrational dread of needles had never left her. As a child she herself had refused to take any intravenous immunizations, having violent fits and convulsions at the sight of a needle coming near her body. Much to her parents' concern, she had taken only those oral immunizations that had been available then.

Marzena's father had been a popular physician in their small town and had taken good care of her dear older brother's deteriorating health and her own psychosomatic spells. She had been barely twelve years old when her brother died, creating a huge void in their small family. The rational part in her head had tried to reason that this had been a freak tragedy. Over the years, with constant support from her parents, who were grieving also, she'd vigorously tried to steady her deep emotional distress and physical repulsion of needle injections, with endless therapy sessions and meditation techniques.

When nothing made visible improvements in her needle phobia, it was at her dad's suggestion that Marzena had decided to pursue further studies in the USA. He had suggested that a change of place would be good for Marzena's mental health and emotional growth. It was partly because she wanted to overcome this persistent crazy needle phobia that she'd decided to go into nursing as her chosen profession. She'd argued with herself that the only way to overcome her violent posttraumatic childhood misconception would be if she opted for a career where she could witness firsthand how injections could cure. And so after some planning and much hard work she had become a pediatric nurse helping acutely sick children, many of them with needle phobia. Mostly her decision had had positive results. Her parents back in Poland were thrilled at her improvements both in her career and in her health. But try as she might, even now she couldn't bear the sight of needles coming close to her own body and did not take flu shots, only oral preventives.

Marzena's illogical dread on seeing the innocuous Space Needle was a first after a very long time. The unpleasant memory made her sprint even faster on the treadmill. Not only her body but her mind too, she was convinced, needed a serious workout. Why had her phobia come back so strangely? Had it never left her at all? Had it just been dormant, waiting for unforeseen stress to raise its sinister head? Actually, she probably had a good idea about the cause of her relapse panic attack on the Needle grounds. It was because the terror in her past had been reflected in present unsettling events back home.

Just before starting her new job last fall, Marzena had visited Gdansk to make arrangements for bringing her parents over to stay with her in upstate New York. But her plans had been rudely shaken at what she found out on her arrival at her childhood home.

Marzena had been expecting her parents to receive her at the airport. So she had been rather disappointed to see that her uncle Patrick, her mother's only surviving brother, had come to pick her up. She had laughingly hugged him joking about his debonair, greying hair. But she hadn't been able to hide her unease at not meeting her parents first.

As Uncle Patrick started talking, she understood why her parents weren't there. On her mother's strict instructions her family had kept the news from Marzena, he began. Her dear mother knew how sensitive her only daughter was and how close she was with her parents. So any long-distance unsettling news about family was sure to take a toll on her graduation studies and job interviewing efforts. Uncle Patrick had ruefully recounted that Marzena's parents had gone through a bitter divorce recently, with her dad leaving her mother for a much younger woman. He had not been monetarily mean though, her uncle had added quickly, but had set up a decent alimony for his ex-wife. Also Marzena had learned, her dad had even set aside some monthly provision for her, his only daughter. Appalled as she had been when she heard about this sudden breakup, it hurt all the more that her parents had kept the news from her for so long.

Was it all right for her father to leave a family of more than twenty years and then pay compensation for the pain? Still in shock, she had recalled that her parents had been unable to attend her graduation ceremony that summer. Her dad had emailed her about being busy with his patients and her mom had made fatigue and other health issues the reason for not traveling. No doubt, these had seemed mere excuses to a disappointed Marzena, but never for a moment could she have guessed that things were not the same anymore between her dear parents. As she had listened to her uncle's halting words, everything went into a meaningless spin before her eyes. The familiar landmarks of her home town whizzing by outside their speeding car had seemed an unrecognizable blur. Could people really move apart so drastically after going through so much together?

When Uncle Patrick told her they were going straight to the downtown main hospital from the airport, she assumed he was taking her to meet her dad who practiced there. She had protested vehemently and had told him curtly that that was not going to happen for she would like to meet her mom first. Uncle

Patrick had remained silent before answering that he knew she would like to meet her mother first. Marzena remembered how puzzled she had been then. Had her mom started working at the hospital too?

Cutting into her distracted thoughts and jet-lagged mind, Uncle Patrick's words had hit her head like a gunshot echoing and reechoing as if she were trapped in a mountain ringed wilderness. He had said in a hushed voice that just before her divorce Marzena's mother had been diagnosed with an advanced stage of breast cancer. Marzena had really broken down then. This cruel disease had not only attacked her mother physically, it had perhaps also dealt a merciless blow to her marriage. She had sobbed uncontrollably.

Marzena's uncle had explained that he was taking her from the airport straight to the hospital, to be with her mom, for she was undergoing chemotherapy there. Right away Marzena was thrown into double jeopardy. On the one hand she had to wrestle with the thought of her mom's acute illness. On the other hand, her mom's treatment involving intense chemo through injections would require needles, those dreaded needles! How Marzena had abhorred them since childhood! By the time they reached her mom's Oncology Care unit, Marzena had started having spasms and had almost collapsed. She had to be strong though for her mother's sake and had encouraged her mother to hang in there and fight. She would always have her daughter by her side.

Marzena's Poland stay with her ailing mother had had to be a short one. She hadn't planned on taking any extended vacation before starting work. Since her parents had not been able to make it to her graduation, her reason for visiting Gdansk had been getting her parents' blessing before starting a new phase in her life. She had also wanted to arrange formalities for them to move to upstate New York, where she would be working. But destiny had other designs and it had been with a heavy heart that she had returned to New York to start her first day at work.

Before long, Marzena found her new boss, Dr. Wong, to be kind and understanding. He reminded her of the dad she had once known. She was able to confide in Dr. Wong about her concern for her mother's health and her own needle phobia. Dr. Wong had given her much hope and information about advanced cancer research, encouraging her to be positive about her mom. He had also made arrangements for her to join a hospital team to attend and participate in this five-day conference about acute needle phobia in younger children and the role of oral medications.

This early-morning hour at the gym was proving to be a real mental catharsis and emotional purging for Marzena. Reliving hurtful memories is sometimes the best way of dealing with them and moving on, thought Marzena, looking reflectively at her sweat-streaked image in the hall mirror, as she exited the gym. She took the elevator to her room, going over the rest of her day's schedule in her head.

3

KAMOLA

Murmur of the River

KAMOLA WENT BY THE RIVER when the western sky was painted in pink and orange hues. Intuitively she knew when this time of the day approached. There was a subtle change in the color of the light which made the trees and the earth look magical. If she paid attention she could tell from the birds' chirping that it was time for them to return to their nests. Like any other day, today also she knew that twilight was approaching. Somehow nature sent her the message.

There, by the river, she could finally be by herself. She did not have to think about finishing any chore for anyone. She could just sit under the big tamarind tree and look at the flowing water in front of her. On its way to the ocean, the river circled her village just like a

Three Daughters, Three Journeys: Quest for Cancer Cure
Ananda M. Chakrabarty, Jill Charles, Indrani Mondal,
and Ranjita Chattopadhyay
Copyright © 2017 Pan Stanford Publishing Pte. Ltd.
ISBN 978-981-4745-90-1 (Hardcover), 978-1-315-19666-4 (eBook)
www.panstanford.com

golden bangle circled the smooth and strong hands of her village women. She had heard the story of the origin of the river from her mother.

A long time ago in a faraway land there lived two sisters. They got married on the same day to two handsome, wealthy and strong brothers when they were really little. So the girls' parents kept them in their parental home for a few more years. Eventually the girls grew taller and stronger. Their parents decided to send them to their husbands' homes. It was a long way. Despite their earnest effort they could not reach their destination before dark. On the way they were attacked by a group of robbers. The servants and guards accompanying these two girls were extremely afraid and ran away. One of the sisters was taken away by the robbers. The other one was able to hide herself. For the first time in their lives the sisters were separated. The sister who was saved from the robbers sat all alone under the big banyan tree which still stood at the entrance of the village. Scared and forlorn, she cried all night. Since it was already dark no villager came to help her even though they heard her cry. The next day, at the crack of dawn, some folks from the village came to offer her help. But she was nowhere to be found. In her place, a small river appeared encircling the village. The villagers named it "Ashru," meaning tears.

Kamola had heard the story innumerable times. She loved how her mother told the story. As a child she was never tired of listening to it. She liked the part where her mother would enact how the girls got ready to go to their in-laws' house, putting on new and colorful saris, gold jewelry, garlands in their hair, *kajal* (a fragrant eye liner) around their eyes. She loved how her mother's tone would change dramatically from being happy to scared to sad. Kamola disliked the part though where her mother would act out foreshadowing the sister who cried all night, alone and helpless. With her small arms, Kamola would give her mother a tight hug every time she reached that part of the story. Kamola still remembered that.

The name of the village was Palampur. It was situated in the western part of Bengal, a medium-size state in India. Most people in the village were farmers. That did not necessarily mean they all owned land. Some worked in the fields of other farmers who were wealthy and owned their own land. Kamola's father and two older brothers, Shibu and Ramen, did that for their livelihood. At present Kamola was the only female in the family. Fragrance from an unknown flower wafted in the air. The breeze was almost enchanting. It was spring time. In fact today was the day before Holi, a festival where people decorate themselves with color of all sorts. Men and women of Palampur went to celebrate this special time of the year. After a day of hard work in the field, they gathered at the clearing adjacent to the big brick building in the village. It was the landlord's house. Tonight he gave special permission to his workers to use his property to celebrate the festivity. Tonight as the moon would rise, big, yellow and almost a full circle, all the villagers, irrespective of their age and gender, would sing and dance. They would get drunk. And they would live in that moment of ecstasy forgetting their day-to-day struggle for living. They would lose themselves completely in the merriment. Kamola's two older brothers and father went to join the others too.

They had a long day as usual. Their day started as the eastern sky was kissed by the crimson glow of the sun. They took care of some household work, took a quick shower, got dressed and sat down almost immediately for a big breakfast of rice soaked in cold water overnight, a piece of lemon wedge, salt, half of a raw onion and a couple of green chilies. As long as her mother was there, she would see that each male had enough to eat. Some days Kamola would wake up as early as her mother. With her eyes still heavy and tired she sat next to her mother and watched. The combination of colors—white rice, purple onion, yellow lime and green chilies—made a lasting impression on her. The foods served were not rare delicacies but they were served with love. There was a sense of contentment, happiness and warmth in the family. That was a few years ago. Now being the only girl in the family at sixteen, she took the responsibility of feeding her father and brothers before they left for the day to work in the paddy field of the village headman. She performed other duties in the household just like a lady of the house would do. She was never forced to take up those responsibilities. She did the chores on her own. It helped her to cope with the loss of her mother. Oh! How she missed her mother!

She could never forget the pain and suffering her mother went through. The local doctor whom everyone saw for fever, cough and colds could not cure her mother's pain in her belly. The year Kamola turned twelve, on some evenings like this, her mother would lay in the courtyard of their mud hut, exhausted and restless. She was not the happy, energetic and beautiful woman anymore. She lost a lot of weight. Her thick, glossy mass of dark hair started thinning remarkably.

As she lay down, she would clutch her tummy and roll around. She was unable to express her discomfort in words. If Kamola or her father would ask her what they could do to ease her pain a little, with the motion of her hands she asked them to go away. This went on for a few months. Her father tried as much as he could to relieve her pain but nothing worked. He even took her to the health center in the nearest town. Somehow Kamola remembered that the doctors thought it was already too late to do anything for her mother. About four years ago, on a spring night like this, she asked Kamola to sit next to her. Kamola did so without any question. She asked where Shibu and Ramen were. It was very hard for her to form those words. Her breath was raspy. Kamola ran to check where her brothers were. And that was the last she saw her mom alive. When she came back with the news of her brothers, their mother had already passed away. Kamola screamed. Like a fanatic, she shook her mother's thin, lifeless body, pleaded her again and again to open her eyes at least one more time. But in vain. She had left the world for good and with her Kamola lost her friend, philosopher and guide.

The memory was still too vivid. It was quiet around her now. The sparrows, the busy crows were not heard any more. The heat died down a little. The leaves of the trees rustled quietly in the southern breeze. Kamola was sitting on a log. This was her time for solitude and calm. She avoided going to the festival tonight. Some girls of her age requested her to join them there. But she refused. She felt no attraction to sing and dance like everyone else. She knew that if she went, she would see Samir, her oldest brother's friend and son of a rich landlord. He was a handsome young man. Unlike other village boys he was attending college somewhere far from the village. He came to visit his family on vacation. He usually treated Kamola with a special kindness. Kamola also developed some sort of liking for Samir. Over the years, she started enjoying his attention toward her. Tonight Kamola did not even want to see him. She was troubled by her own worries.

There was a legend in her village. If someone was sad and worried for some unexplainable reason and came by the river "Ashru," in its murmur against the stones, that person would hear some kind of guidance leading to the path of joy and happiness. Kamola just wished that her mother had not died because of some pain in her belly that nobody could stop. If her mother was here, she could just tell her about her worries and things that had been bothering her recently. But she felt all alone. Once again she cried like she did four years ago sitting next to her mother's lifeless body. The sound of her sobbing and the river's murmur were intermingled.

4

SELENA

The Test

"There's a lump," said Dr. Weiss.

Selena flinched, but said nothing as her doctor took her hand. Dr. Weiss gently guided her hand to the spot between the breast and armpit on Selena's body. She pressed Selena's fingers softly against the skin. *I would never have noticed it*, Selena thought, *but here it is; my body has betrayed me.*

Dr. Weiss told her, "It's small, just the size of a grain of rice, but you should have it checked, especially with your family history."

The doctor leaned back on her stool and wiped her glasses on the hem of her lab coat. Sunshine filtered through the blinds and shone on her pink cheeks and wavy silver hair.

Three Daughters, Three Journeys: Quest for Cancer Cure
Ananda M. Chakrabarty, Jill Charles, Indrani Mondal,
and Ranjita Chattopadhyay
Copyright © 2017 Pan Stanford Publishing Pte. Ltd.
ISBN 978-981-4745-90-1 (Hardcover), 978-1-315-19666-4 (eBook)
www.panstanford.com

Selena felt the lump again to be sure. Could her doctor have made a mistake? No. It was too small to see but she felt it there. Selena sat up, climbed off the examining table and began to get dressed again in her own clothes. Hospital gowns always made her feel cold and overexposed.

Selena had lost her Mama and Grandma Lily to breast and ovarian cancer. Four aunts, three of her mother's sisters and her father's one sister had all been diagnosed with breast cancer after turning sixty.

"I knew this might happen to me, but I didn't think it would be for a long time," Selena said.

Dr. Weiss' blue eyes reflected her concern for Selena; she had watched her patient grow up after losing her mother.

"It's very likely a benign lump, but I'd like you to have an MRI of the breasts and ovaries, just to be certain. Nazreen can help you set the appointment."

"Why an MRI rather than a mammogram?"

"I want to avoid the small amount of radiation in the mammogram and the MRI can check both breasts and both ovaries. A mammogram or an ultrasound reading might miss something, but an MRI is the most thorough examination and with sound waves and no radiation."

Selena nodded and said, "I'll set it up. Thank you."

Selena trusted Dr. Isabel Weiss completely and had seen her for all her appointments since she turned twelve and was too old for a pediatrician. As a gynecologist and surgeon, Dr. Weiss had treated many women for cancer, from diagnosis to lumpectomy or mastectomy or hysterectomy. She worked with a network of oncologists to medicate and monitor patients to prevent a recurrence of cancer and with counselors to help patients understand and prepare for treatment, to face their fears with as much knowledge as possible. Dr. Weiss helped Selena find a therapist after her mother's death. When Selena felt too depressed to get out of bed and go to her classes, Dr. Weiss convinced her that trying counseling and antidepressants would lift the unbearable weight of her grief.

"No matter what happens in your body, you'll always be you." Dr. Weiss had reassured her then.

Selena had needed a year of antidepressants and counseling but she had excelled at the University of Chicago and chosen psychology as her major. She wanted to understand human minds, not only through lab experiments and statistical research but by listening to others who had lost a loved one to death or divorce, or struggled to treat their depression or mood disorders and find an independent, healthy life. Now twenty-five years old, Selena was in her third year of a doctoral program in psychology, teaching introductory courses to undergraduates and volunteering as a counselor at a homeless shelter and a conflict mediator at a high school that could not afford counselors. Every time she helped someone else rise out of her suffering, Selena felt that her losses were not in vain, that she was still close to her mother, caring like Mama had taken care of her.

Dr. Weiss had set a good example as a care provider who looked out for her patients, helping them find specialists in physical and mental health care when needed. Selena had heard Dr. Weiss argue with an insurance company on the phone to make sure a patient's mammogram was paid for. Health mattered more to her than income and she continued to accept uninsured and unemployed patients and many with public healthcare, as well as patients like Selena, who had always been insured through her father.

Just before Selena left the examining room, Dr. Weiss asked, "Have you thought any more about getting a genetic test?"

"No one in my family ever had one, except Mama," said Selena.

She thought of her eight female cousins as well as herself. Clarice had just gotten married. Marta had two sons now. The youngest cousin, Kiara, just started third grade this fall. Any one of her cousins could die early from breast or ovarian cancer.

"I don't know if I want to know," Selena said. "And I have doubts about giving money to a testing company that tried to patent the BRCA1 and BRCA2 gene. Nobody should own any part of another human's body; America got rid of that in 1863."

"I understand that," said Dr. Weiss. "It's frustrating that genetic tests cost so much and aren't available to every patient with a family history of cancer. It's a risk for women and for men who can get prostate cancer from the BRCA1 and BRCA2 mutations. There are many genetic causes of cancer yet to be discovered. If you don't think you could live with knowing if you have the genetic mutation, it is better not to have the test done. For a patient who does carry both mutations, a preventative mastectomy and removal of ovaries can reduce the risk of developing cancer by 90 percent. I have had patients who have their ovaries removed, or their breasts, or both. Breast reconstruction is an option too."

Selena had read up on preventative surgeries and the side effects. She worried about early menopause and the thought of having any body part removed frightened her. Even if she could have her breasts reconstructed from belly fat, even if her skin surface looked the same except for a few scars, would she *feel* the same? She lived comfortably in her body; her periods never hurt her or made her moody. She had hoped to have a child some day, years from now after she had married her boyfriend Claude Tate, finished her PhD and started a counseling practice.

"Mama lost both breasts and both ovaries and died anyway." Selena said sadly.

"Your mother was in stage III when her cancer was diagnosed," Dr. Weiss said. "I know you're worried about the lump, which is reasonable. The chances of it being benign are about 90 percent and even if it contains cancerous cells, it would only be a single stage I tumor. The survival rate for stage I is 90 percent."

"Which is why you're obsessed with early detection," said Selena. "You take good care of your patients. I appreciate you looking out for me and I know how lucky I am to have the choice of preventative care. Some of the women I counsel at the shelter have never been to a doctor, much less had a mammogram."

"If they need one, you give them my card," said Dr. Weiss. "Take care of yourself, Selena and say hi to your dad. I'll call you as soon as we have the MRI results."

Selena thought of the genetic test as she went back to work. She ruminated on it as she hurried between ivy-covered gray stone lecture halls of the University of Chicago to her classes, as she discussed the symptoms of depression with her Psychology 101 students on, as she washed her hands in the women's restroom under the chart of "How to Do Your Monthly Breast Self-Exam."

The next day she kept thinking about it as she hugged a crying high school senior whose father had died of a heart attack. She thought of it in the women's shelter as she acted out a job interview with a laid-off sixty-three-year-old homeless lady determined to "get back to work in accounting where I belong."

Selena's health worries seemed minor compared to others who fought for survival daily. She had her work and classes at the University of Chicago. She had her dear dad, Grandma Grace, her best friend Janine and so many others and Claude, who loved her. She had her quiet Hyde Park apartment, full of books and plants and her dog Jojo, still wagging her tail at the door every time she returned home. Selena had never had to wonder where her next meal would come from, where she would sleep at night, whether she could afford medicine if she got sick. Maybe it was braver to take the genetic test. What a relief it could be for her if she didn't have the dangerous mutations! And if she did, wasn't it better to try to prevent the cancer now? If she had the test, she felt obligated to tell her relatives about the results either way. They deserved to know.

When Selena returned home after a busy day, she dragged herself up three floors of stairs in her old brick apartment building. She could hear Jojo rushing to the door as she unlocked it. Jojo wagged her tail and opened her mouth in a doggy grin as Selena petted her. Grandma Grace came over to walk Jojo in the afternoons because Selena got home so late, but Jojo always felt delighted to see her. The black Greyhound was ten years old now, but still enthusiastic as ever, eager to walk or jog by Lake Michigan with Selena leading her on a leash.

"It's good to see you too, girl," she said. "You're the only one who loves me without ever worrying about me."

The apartment had seemed lonelier since Janine moved out in June to get married. Selena's boyfriend Claude was eager to live with her but she questioned whether an engagement should come first. They had been dating for three years now and she knew all his eating, sleeping and housework habits (or lack thereof) without living with him.

Claude had met Selena in an undergraduate psychology class. They loved to swim at the Chicago lake beaches in summer, to watch late-night movies together and to go dancing at the Blue Jay. Claude worked as a computer systems administrator for a shipping company and earned a lot more than Selena did as a graduate student.

"We can save up and buy our own house," Claude had assured her. "We could do that a lot quicker if I move in with you."

His own apartment was a South Loop studio on the 37th floor with an amazing view of downtown and Lake Michigan and most surfaces covered in old socks, dirty cups and computer games.

On her cell phone, Selena noticed messages from Janine and Claude. She wondered who to tell first about her lump. She resolved not to tell her father anything until she had the MRI results and knew if she had cancer or not. Daddy and Grandma Grace had suffered enough losing Mama and Selena would never worry them about her in any way if she could avoid it. Still, she needed to share her worries with someone.

Jojo's tail thumped against the wall of the narrow hallway as Selena filled her water and food bowls. The yellow and green leaves of the maple trees rustled outside and the September sky shone bright blue. Selena would go to the place that always helped her sort out her feelings.

She fastened Jojo's leash to her collar and said, "Let's go, girl."

They walked together down the tree-lined streets of Hyde Park, past Art Deco apartments and cafes offering tasty Italian espresso, Middle Eastern falafel and spicy Thai noodles. They passed grade schools, church spires and the El tracks with the silver train rushing over tracks two stories above the street. Selena and Jojo walked fast, only slowing down for crosswalk lights and to let a little boy walking with his mom pet Jojo.

At last they arrived at the lake at 55th Street Beach. Beyond the tan sands of the beach and crashing white waves, Lake Michigan shone blue and infinite as the sky. Here Selena felt connected to the whole world, to all nature, to all the people she loved, living and dead, and to God and the universe. The world was huge and beautiful and her problem, any problem, was temporary. She was not alone, ever, as Mama had made her promise to remember. Selena could face her fears, do her best to protect herself from cancer and help her cousins and her clients and students too. Jojo pulled her onto the lake trail and they jogged together for miles, past the rustling oaks and maples slowly turning red and gold, past the blue waves and white sailboats, past sandy beaches and runners and bikers, children building sandcastles, not to last, but so beautiful to work on.

Miles later and hours later, after looping around and returning to Hyde Park, Selena called Janine.

"Hi Selena. What's up?"

"Are you busy? I need to talk, in person, if you have time."

"I'm just cleaning house, but that can wait. Where should I meet you?"

"55th Street Beach."

5

MARZENA

Day

A FEW HOURS LATER a young lady, Marzena Sumak, a nurse from the Neonatology Department at Central Hospital of Upstate New York, was presenting a paper on "Symptoms and Cures of Needle Phobia in younger children." The large hall of Starr Conference Center was quite full. The audience listened intently to the tall brunette dressed in sharp black and white professional attire, as she explained the possible causes of and methods for battling the often crippling psychosomatic effects of needle phobia. She emphasized how oral medications would be the best curative procedure in such cases. There was a loud cheer when she ended and the moderator announced the beginning of the Q and A session.

Three Daughters, Three Journeys: Quest for Cancer Cure
Ananda M. Chakrabarty, Jill Charles, Indrani Mondal,
and Ranjita Chattopadhyay
Copyright © 2017 Pan Stanford Publishing Pte. Ltd.
ISBN 978-981-4745-90-1 (Hardcover), 978-1-315-19666-4 (eBook)
www.panstanford.com

Almost instantly a hand shot up from the third row and when asked to speak, a bespectacled, buff young man with a head of tousled dark blonde hair stood up. He took the mic and introduced himself as Dr. Eryk Cyrek, research fellow at Sanchez Cancer Institute in Central America.

Eryk Cyrek? The Eryk Cyrek, heartthrob of Marzena's high school days in Gdansk? But what was he doing here and why did he look so different? No, this was surely a different Eryk! So disconcerted was Marzena that she didn't quite hear the first part of his question at all. She was busy staring at him trying to compare this unkempt, heavy built, research scientist with the lithe senior, varsity athletic champ from her freshman year in high school. He was commenting, in a deep European accent (at least that accent seemed familiar) that even though not very prevalent yet a very grave concern, in fact the greatest challenge of needle phobia, was finding oral cures in serious medical diseases like cancer. For example, he asked, did the presenter know any method through which chemotherapy by intravenous shots could be replaced by equally effective, potent, cancer curative oral medications? Was medical science advancing enough to answer this challenge? Could medical science provide alternative oral treatments for curing such acutely ill children and adults? As Marzena was coming up with something to say, the silence in the hall following the question was broken by a buzz of voices.

Eryk Cyrek waited an instant and then continued with gusto that he had been invited to this conference en route to his hometown in Poland to present some research papers and work on clinical trials in some hospitals there to explore the efficacy of one such alternative cancer cure without the usual chemo. He said, in collaboration with other medical researchers, he was trying to change the face of conventional medicine.

He went on in a genial mood. It had all started several years ago when Eryk had been doing his residency in Oncology in Midwestern Clinic in the central US. There at a session of the international conference on "New Horizons in Cancer Cures," he had first met and heard Dr. Sam Roy from Blue Frost Land Research Foundation, talking about a protein that he had extracted from marine bacteria. Dr. Roy had discovered this protein as having breast and ovarian cancer curing properties, among others. Roy had named this small protein Neelazin. Over the years Dr. Roy had been able to show amazing positive results in cancer treatment when this Neelazin was administered as an intravenous injection to his advanced cancer patients.

That inspiring presentation had set Eryk and a few of his colleagues thinking and during the after-conference dinner with Dr. Roy, Eryk had had a new vision. Completing his MD PhD, Eryk had moved to Sanchez Research Institute in Central America as a Senior Research Scientist. He had collaborated with Dr. Sam Roy few times and working hard with his new research group in Central America, Eryk had tried to push Dr. Sam Roy's research forward to meet further challenges in the medical field. Eryk's first task had been to retest Neelazin. Needless to say, he and his fellow researchers got awesome cancer curing results with this Neelazin protein. They also found out that this protein was a tad larger than insulin. Finally Eryk's research on modifying the Neelazin with enteric coatings and microemulsions containing surfactants and fatty acids went on to show that this modified Neelazin could cross the gut intestinal layers and be taken orally with equally viable curative and preventive results for malignant tumors.

Shaken as she was by running into her romantic interest from high school, Marzena couldn't help countering Eryk instantly. "I think many of us here have read in medical journals about the phenomenal results of Dr. Sam Roy's research medicines. But interesting as it sounds, forgive me for saying, that it is your contention, Dr. Cyrek, that totally confounds me, let me correct that, confounds all of us present here," she broke in as Eryk stopped to take in the hushed excitement in the conference audience. "You just mentioned that this protein you're talking about, is larger than insulin. Did you not?" she continued. "As far as I know from my experience as a nurse, insulin, in spite of being smaller than your protein, still has to be taken intravenously, even though researchers have been trying for a long time to come up with an oral insulin."

Eryk laughed heartily and said, "Of course I know you're a nurse and therefore quite knowledgeable about drugs and their mode of administration. Actually I think most attendees of this conference are knowledgeable about drug administrations. So I'm sure you'll not mind if I give you all a little professional jargon, as they say." He continued with empathy, "During my years of residency I treated several diabetic patients who were always complaining, and with reason, about having to take painful insulin shots several times daily to keep their blood sugar level in check. Many of them hated the sight of needles but had to endure this daily torture to stay healthy." Then turning to Marzena he said energetically, "You're right of course that there is no oral insulin in the market yet. But, mark my words, a lot of progress is being made in that direction. You're also right that a protein and its peptide fragments are large enough in size, ensuring their low absorption through the gut as well as enzymatic degradation in the gut. So they are not orally effective."

Marzena countered with spirit as voices rose from the audience. "That's why small molecule drugs have an advantage. Only these can be taken orally because they can pass through the intestinal lumen with ease, just my point. So how were you able to make this larger protein orally effective, Dr. Cyrek?" she challenged.

Eryk ran his hands through his disheveled hair impatiently. "Allow me to explain please" he began, addressing the audience with a dramatic sweep of his free hand. "That's precisely the point of our research. Many of us know why small molecule drugs are effective when taken orally. The reason is they can pass easily through the inner layers of the intestine. The epithelial cells here are very compact, with tight junctions. And this is also what prevents the passage of larger molecules. But do let me add," he said emphatically, "Much progress, yes, I mean, much progress," Eryk paused and gestured again with his expressive hands, "is being made to come up with a modified insulin with additives like enteric coatings or with chemical modifications that significantly enhance their absorption through the intestine. No doubt there still remains the problem of the right amount of oral insulin consumption to prevent risk of hypoglycemia or dangerously low blood sugar in diabetic patients." He cleared his throat briefly and threw a question at the general audience. "Since most of you are involved in health care, aren't any of you aware that a large multinational company has just recently announced an oral peptide drug for lowering blood glucose but involving low risk of hypoglycemia in diabetic patients? In fact this drug will be tested very soon in a large clinical trial in about eight thousand patients on two different continents."

Marzena cut into what she thought were Eryk's straying comments. "But as far as I remember you were telling us how an oral anti-cancer protein drug was possible?"

"Well I was just giving a preamble to better explain what my research team and I have been doing all these years," Eryk smiled almost apologetically. "Neelazin as many of us know works best with curing breast and ovarian cancers through intravenous injections requiring needles." Eryk raised his expressive hands again in a grand gesture and continued.

"I have used the same technique of chemical modification that the insulin people are using, to come up with an oral form of this Neelazin. And," he paused for effect, took off his glasses and dropped his voice to almost a whisper, "I just found out that it works! It works!" He raised his arms up triumphantly.

There was pin drop silence in the Conference Hall. At this point the moderator announced that he was closing this session as it was time for the next presentation. Marzena's session ended amidst excited questions, deep interest and muffled disbelief.

6

KAMOLA

The Magic World

KAMOLA'S MOTHER, NIRU, WAS quite young when she got married to Kamola's father. Niru lost her mother when she was a mere child. Kamola heard from her mother that no one could figure out what was happening to her. Niru grew up strong and beautiful even without her mother's guidance. She was a good wife and a good mother. Her presence was like the steady lamplight in their mud hut. Whoever came in touch with her basked in the glow of her warm kindness. She did all the hard work in a poor peasant's family without any complaints. Kamola remembered every morning after taking a shower, her mother would dry her thick, long hair in a cotton towel and put *sindoor* (vermillion powder) on her forehead. The entire process would not take more

Three Daughters, Three Journeys: Quest for Cancer Cure
Ananda M. Chakrabarty, Jill Charles, Indrani Mondal,
and Ranjita Chattopadhyay
Copyright © 2017 Pan Stanford Publishing Pte. Ltd.
ISBN 978-981-4745-90-1 (Hardcover), 978-1-315-19666-4 (eBook)
www.panstanford.com

than ten minutes. As a child, Kamola used to stare at her mother getting ready mesmerized. She thought at that moment her familiar ma was transformed into a goddess who could kill all the demons and protect her children.

As she grew older she started adoring her mother just as a beautiful woman, inside and out. In the evening her mother would call her inside the house. She would ask Kamola to sit cross-legged on the floor. Her mother would kneel behind her. Then very gently and affectionately she would undo Kamola's hair.

"Kamli"—that was what her ma used to call her—"how did you get your hair so tangled up?"

Kamola did not answer. She almost fell asleep as her mother gently massaged slightly warm coconut oil through her knotted hair.

"Have you been climbing the mango tree with those boys? I told you so many times not to do so." She scolded her in fake anger.

At this Kamola laughed out loud. She was tall and thin for her age. That was her big advantage over those boys. She could climb any tree faster than her playmates. Being frustrated they would try to slow her down by pulling on her long braids. Yes! Her hair was manhandled all the time. But she did not care. She used to run around with Shibu and Ramen and their friends as soon as they had some free time. For some reason Kamola was never close to any girl of her age. Thinking all that Kamola closed her eyes for a moment. Still now she could feel her mother's touch on her hair. How could she leave her? How could Kamola's Goddess Mother leave her only girl behind? Where was she when she needed her so much?

Nothing was the same in Kamola's world any more. She was no longer a skinny, mischievous, carefree girl racing boys and climbing mango trees. Her playmates all grew bigger and taller just like her own siblings. Some of them even stopped talking to her. As if by some unwritten laws of nature, they knew that they belonged to two different worlds. Kamola's brothers started helping out their father more in the fields. That left them very little time to play or cause mischief. Kamola rarely saw her brothers around these days. She missed their company too.

In the past on certain rainy nights, all of them would gather around their father to listen to ghost stories in the dim light of a candle. Their father was an amazing story teller. He could not act the stories out like their mother. But he was a magician with his voice. With all the drama in his voice, he could bring any story into life. Even though he was tired from doing hard work from dawn to dusk, he could not deny the children's request for telling stories. The brothers and sister held each other tight and close as the suspense was built up in those stories. By the time their father finished, dinner would be ready. No one wanted to let go of each other because they were still scared of those imaginary ghosts of their father's stories. But it was too hard to resist the temptation of hot *khichuri* (a porridge-like dish made out of rice, lentils and spices) served with melted butter, a rare delicacy in the household.

After it stopped pouring, the next morning there would be small streams of water running on the village streets. Shibu and Ramen would manage to grab a cut up banana stem and float it in the flowing water. She would want to join in but they would not let her, saying that she would make the banana stem boat sink. Kamola used to get mad at them and planned to get them in trouble. Now all those happy memories made her pensive.

7

SELENA

Why Right Now?

JANINE MET SELENA AT 55th Street Beach, carrying blueberry scones and their favorite coffees: Selena's soy cappuccino and Janine's orange mocha. Janine didn't know what was wrong but showed up prepared to take care of her best friend, regardless. The two friends sat on a big gray rock and looked out at Lake Michigan. She asked Selena what was wrong and listened patiently to her worries about the lump in her left breast and her decision to get a genetic test after her MRI.

"I thought you never wanted a genetic test," Janine said. "You said it was a rip-off for patients and you would never have any healthy body part removed."

"I've read up on it and I do think it's better to know if I have the genetic mutations or not. When Dr. Weiss

Three Daughters, Three Journeys: Quest for Cancer Cure
Ananda M. Chakrabarty, Jill Charles, Indrani Mondal,
and Ranjita Chattopadhyay
Copyright © 2017 Pan Stanford Publishing Pte. Ltd.
ISBN 978-981-4745-90-1 (Hardcover), 978-1-315-19666-4 (eBook)
www.panstanford.com

found the lump, it seemed so much more real to me. I always knew I could get cancer like Mama. I never deluded myself that I was immortal but I did hope it would come later, after I had a chance to start my career, start a family. Most women in my family who got cancer got it in their sixties and Uncle Roger didn't get prostate cancer until his eighties, but Mama and Grandma Lily got cancer in their forties. Now I don't know if I'm going to see eighty or even sixty. I feel like I need to know if I have this mutated gene and I need to tell my aunts and cousins about it too. It ran in Mama's family and maybe in Daddy's. I need to warn them."

"I totally get that," Janine said. "It's your decision and I support you, whatever you choose. If you get the MRI and you do have cancer, what would your treatment be?"

"I would have a double mastectomy and an oovectomy, have my ovaries taken out, because estrogen feeds this type of cancer. I would have chemotherapy and maybe radiation and medicine after that."

Janine sipped her coffee and turned from the lake to look Selena in the eye.

"But the lump may not even be breast cancer. Dr. Weiss said there's a 90 percent chance it's not."

Selena petted Jojo, who lay at their feet, uncharacteristically quiet now.

"I think my family is the 10 percent," Selena said. "Even if I don't have cancer now, I want to have the genetic test done. And if it comes back positive, I need to protect myself, not wait until I actually have cancer. I would get a mastectomy and an oovectomy if I have the BRCA mutations."

Janine's eyes widened and she asked, "Are you sure? Would you want to wait until you had a baby?"

Selena shook her head.

"I always thought about motherhood as something I could do, would do, but I never wanted it now, always after college, after grad school, absolutely after I had gotten married and been married for a few years. I couldn't raise a child alone. Some parents are strong enough for that but I wouldn't be. I can see now why Mama and Daddy only had one child; it takes so much time, so much energy."

"You have a ton of energy," Janine said. "And you dedicate yourself so much to your classes, your students, the high school and the women's shelter. You eat Sunday dinner with your dad and Grandma and call them all the time. You always come to my comics panels and graphic novel readings even when you and Eli are the only ones there."

Selena smiled as she thought of Janine's comic book *Weird Girls* about their lives, the funny scenes where they turned into superheroes, the character disguised as a robot who turned out to be human, based on Eli, then Janine's boyfriend, now her husband.

"You're the first one I've told about any of this," Selena said. "Please don't say anything to anyone, not even to Eli until after the MRI. I don't want Dad to know until I have those test results."

"OK," said Janine. "Although Eli could handle it. He's seen all kinds of medical things as a pharmacist and he doesn't gossip. I'll tell him after your MRI, though."

Selena parked her small silver electric car outside Claude's fifty-story high-rise. The steel and glass apartment tower looked imposing, but Selena knew the entry code to the front door and all of the security guards and doormen, by name.

"Hi, Lazaro," she said as she signed in.

Riding the elevator to Claude's 37th floor apartment, she felt her stomach sinking. She knocked on his door, as always, although she had her own key.

"Come on in," Claude called. "I'm barbecuing."

As Selena unlocked the door, she stepped into the reassuring brown sugar, honey and onion smell of Claude's barbecued ribs. Even at her most stressful times, she still got hungry.

"Hi baby!" Claude called from the balcony. "Come on in and help yourself to a drink."

He stood outside, turning the ribs on the grill beside foil-wrapped ears of corn and potatoes and a basket of glistening zucchini, red peppers and asparagus above the blue gas flames.

Claude stood tall and strong in his jeans and T-shirt, with copper brown skin and short black hair. He always changed out of his buttoned-down work shirts and slacks as soon as he got home. He smiled at Selena, his brown eyes sparkling behind black-framed glasses. She hated to hurt him, even to worry him about what might happen to her.

Instantly Claude asked "What's wrong, Selena?"

"I got some bad news today," she said, stepping out onto the balcony.

The city rushed on below them, a maze of honking cars, trucks and cabs, buses and El trains oblivious to individual lives.

Claude turned off the grill and stepped toward Selena.

"The doctor found a lump in my breast," she said.

Claude wrapped his thick arms around her and she buried her face in his chest and cried.

"I love you," Claude said. "I'll do anything to help you. What did Dr. Weiss say?"

"She said it's probably benign. Ninety percent of breast lumps aren't cancerous, but I should get an MRI because of my family history."

Claude stroked Selena's thin black braids and tucked one behind her ear. He kissed her forehead.

"That makes sense," he said. "I know it's hard not to panic but we really shouldn't worry too much about this until you get the MRI results. When will you have it?"

"In two weeks, on the 20th. And you're right, I'm trying not to panic. I don't want to tell Dad or anyone else until I have the test results. I only told Janine and you."

"I won't say anything to your dad or my folks until you do," Claude said. "And if it's benign we don't have to worry them at all."

He kissed Selena again and switched the grill back on.

"I'm getting you a drink," he said. "I insist."

He opened his fridge to offer her a pilsner, white wine, pineapple or orange juice or lemonade or the lime margarita mix he only bought for her.

"Just lemonade, please," she said. "If I drink any alcohol, I'll start crying and I won't stop."

"Okay," Claude said. "But sit out on the balcony, put your feet up and don't worry about anything. I'm going to take care of you, Selena, no matter what."

Selena settled into a lounge chair by the big red grill as Claude handed her a tall glass of lemonade with lemon and lime slices.

"You're so good to me," she said. "How was your day, Claude?"

"The usual fools making the usual mistakes. I had to go in an hour early to help the director redo his presentation at the last minute and then had to flip through the slides for him because he couldn't figure out how."

Selena let her problems slide to the back of her mind listening to Claude's smooth baritone voice, his jokes about work, running to catch the El train and finding a rusty truck parked in his parking space in the garage. He always listened to her patiently, then cheered her up. By the end of dinner she felt almost normal again.

* * *

Two weeks later Selena had her MRI. She had read about MRIs and heard about them from some of her clients, but nothing truly prepares a person for lying still inside a cool metal tube for half an hour, wrapped in hospital blankets and eerie light.

"Thank God I'm not claustrophobic," Selena told the technician, just before lying down on the platform.

"And thank God you have music," said the technician, a kind middle-aged woman, who offered her headphones with choice of any music she liked. Selena asked for blues. Usually she preferred rock and rap but now Daddy and Grandma's music consoled her.

She could lie silent and still hearing Ma, Rainey belt out her pain, "I've got the downhearted blues and I'm in misery."

When the platform slid out of the tube, Selena sighed with relief. The metallic taste of the dye clung to her mouth, although the IV tube had been in her hand, to spread the dye to her breasts. The technician offered her a glass of water, which she gulped.

"You'll get a call with the test results in two weeks," the technician said.

When Dr. Weiss called Selena with the results instead of making an appointment to see her, she supposed that the test results would not be what she dreaded.

"I have good news," said Dr. Weiss. "The breast lump is benign, not a cancer, not even a tumor, only fatty tissue."

"The test is negative?" Selena asked.

"What a relief, right?" said Dr. Weiss. "I hated to worry you but we had to be sure."

"Of course. It's better to know. Dr. Weiss, there's something else."

"Yes?"

"I'd like to have the genetic test done."

"All right. It will take at least a month to get the test results and will require two ultrasounds, of the breasts and the ovaries and uterus."

"I thought it was a blood test."

"The insurance company insists on both ultrasounds before the blood test. Everyone who has the genetic test has to be checked for breast cancer and ovarian cancer first. Even though you just had the MRI, you would need the ultrasounds too. It's excessive, I know."

Selena decided not to tell Daddy, Grandma Grace or anyone in her family about the MRI or the genetic testing until she had the final results. There was no reason now to tell them about a benign lump now.

The ultrasounds were painless, just cold gel and a metal wand gliding over Selena's skin. She resented having to miss her psychology and statistics lectures and refused to miss any hours of the Psych 101 class she taught or her counseling sessions at the high school and the women's shelter.

"I bet you don't even have the stupid mutant genes," Claude assured her. "Even if you do, by the time you're old enough to get cancer, they'll probably have a cure for breast cancer, just a pill you can take instead of chemo."

Selena frowned.

"Anyone is old enough to get cancer. My mom was only forty-one."

"And you're only twenty-five, honey. Stop worrying."

Claude and Selena were at her apartment, snuggled under a blanket on the couch with Jojo at their feet. Selena braced herself to tell him the thing she'd been dreading.

"Claude, if that genetic test comes back positive, I will have preventative surgery. I will have a mastectomy and an oovectomy."

"Oovectomy? You'd have your ovaries taken out? Why? I mean why now before we even have a chance to have children?"

"I don't want to wait until I have cancer. I don't want to get a mammogram every year hoping and praying that I'll be healthy this time."

Claude protested "Most women don't even have a mammogram until they turn fifty."

"Mom and Grandma Lily were dead by fifty. I'm not waiting."

Claude hugged Selena, to comfort her or to hang onto her, she could not tell which.

"I know you don't want to get cancer like your mom, but honey, she put off getting treatment until it was too late."

"So I won't."

"But why right now? Why when you're twenty-five and healthy? There could be better cures in twenty years."

"I may not have twenty years."

"You might not get cancer until you're in your sixties like your aunts, or maybe never at all. Why ruin your body while you're still young?"

"Why let my body die young because of vanity?!"

Selena kicked off the blanket and stood up.

Claude stood up too and began to pace the room. Jojo whined nervously on the floor like a child watching parents fight. She hopped up on the sofa behind Selena.

"I know you could have reconstructive surgery," Claude said. "It's your body and your final decision. Even if you had cancer right now, I wouldn't leave you, I'd pick you up and carry you after chemo if you needed my help because I know you'd do the same thing for me."

"I would," said Selena. "I would have done a lot more for Mama if she'd let me."

Claude ran his hands nervously over his short hair.

"But Selena, if you have your ovaries removed that doesn't just change your body. That changes our entire future. Maybe you can live with menopause at twenty-five and never having a child, but I can't. I don't want to risk the money and time to freeze your eggs when you won't even have any of the hormones to make you a mother."

"Make me a mother? Wouldn't adoption make me a mother? I don't have to give birth to a child to love them."

Selena faced Claude, inches away but further apart than they'd ever been.

"How can you spring this on me?" Claude demanded. "I want a baby, our baby, part of your family and mine, not some stranger. We have a plan. We'll get engaged and move in together this year, get married next year and have a baby the year after that. Couldn't you wait three years? Couldn't you give us a chance to have one child before you throw your entire healthy reproductive system away?"

Selena quieted, took a deep breath, then said, "You don't get it, do you? You want your baby more than you want me. I have a genetic flaw that can kill my daughters or my sons. I have every right to decide not to pass that gene on. My DNA isn't the only thing I have to give this world."

"You don't want a child like I do," Claude said.

"You've never even changed a diaper. You wouldn't be willing to cook or clean or stay home part-time yet you assume I'll do it all for you. No, I don't want a baby like that."

"Could you really live your whole life without giving birth and not miss it?"

"Yes. I'd be willing to adopt a child or even to have your child with a surrogate mother, but if you only want a child because you expect it to be like you, that's the wrong reason to bring another person into the world."

"You're crazy!" said Claude. "I'd love our baby and take care of our baby and you, Selena. I'd never quit you but I can't give up on being a father. And I wouldn't have a child with a surrogate mother, that's just sick."

"What if we married and I turned out to be infertile? Would you just divorce me?"

Claude looked down and blinked, trying not to cry.

"You're being hysterical, Selena. You're so worried about what might happen in the future that you're letting it wreck the present. Please don't decide this now when you just got the test results. You haven't even dealt with the shock of that or told your family. Please wait until we can marry, Selena. Please don't give up on us and our child."

Claude clung to Selena but she could already feel herself slipping away from him.

8

MARZENA

Twilight

MARZENA WAS RESTLESS WHEN she left the room both because she had run into Eryk so unexpectedly and of course for the revolutionary ideas that he had shared. Gone was the dandy senior who had all the high school girls swooning after him. But then she had last seen him when she was a sophomore and he had graduated with flying colors from their high school with a full scholarship to an Ivy League university in the US. This new avatar of Eryk seemed kind, caring, committed and had actually given her new hope in battling her mother's deadly disease. Could she then dare to believe that now he could actually give her mother a second chance even without the cruel chemo? She would have to get in touch with Eryk as soon as possible.

Three Daughters, Three Journeys: Quest for Cancer Cure
Ananda M. Chakrabarty, Jill Charles, Indrani Mondal,
and Ranjita Chattopadhyay
Copyright © 2017 Pan Stanford Publishing Pte. Ltd.
ISBN 978-981-4745-90-1 (Hardcover), 978-1-315-19666-4 (eBook)
www.panstanford.com

With this thought in mind she walked out stopping for a minute to check her phone. She found that she had several missed calls and equally as many text messages from her uncle. Knowing full well that it was late at night in Gdansk, she quickly called back and waited with a racing heart, unsure of what she'd hear. Uncle Patrick picked up almost instantly and said quietly, "Marzena dear, your dad thinks you should come over right away to see your mom…"

Marzena's instant reaction was angry disbelief. Why didn't her dad call her himself if he thought she should go over? Surely because she hadn't met him on her visit home last summer and had been steadily declining the monthly allowances he sent her, he was trying to bring her home using her mom as an excuse! In fact just last night Marzena had talked with her mom for quite a long time and she seemed to be recovering after her chemo sessions. Marzena simply couldn't figure out any more what to believe or whom to trust. She wanted to explain all this to her uncle on the phone that her world was rudely shaken and she needed time to stabilize. But all she managed to do was cut into her uncle's halting words and say that she was attending a conference in the West Coast. For time off she could only talk with her boss after she returned to work on the East Coast next week. Then she ended the call for she didn't want to raise her voice and argue with her dear uncle.

Marzena couldn't get Eryk Cyrek out of her mind. Images of her idyllic freshman year adoring the blue-eyed Nordic hero flashed through her mind. The senior had been very popular not only in the student circle, but coaches and teachers also referred to him as an example. As a shy freshman herself she had never really been able to approach him closely enough to tell him how she felt about him. He had always had a huge fan following which made her think that he would be quite vain and overbearing. They say distance lends enchantment to the view. Seeing him suddenly after so many years, with her mind unstable due to family complications, Marzena's past adulation for Eryk seemed to blossom instantly into a full-blown longing. Not only was it for the golden boy whom she had had a crush on but hardly knew, it was also a nostalgic ache for the simple joys of her girlhood days. She reasoned, that unlike before, given a chance now, as a qualified, confident and attractive neonatology nurse, Marzena could surely make a favorable impression on Eryk, the research scholar. Most importantly, she wanted to know more about the alternative oral cancer cure he had alluded to in commenting on her paper.

As Eryk was nowhere to be seen in the Starbucks lounge where the conference members usually met in between sessions, Marzena made it a point to read the synopsis of his paper and also attend his presentation later that day. Though very informative, giving Marzena another way of thinking about her mother's recovery, she caught herself repeatedly trying to make sure, this was really her high school heartthrob. She kept looking for a chance to catch Eryk by himself after his presentation to find out if he remembered her at all.

Throughout Eryk's presentation she couldn't stop thinking about her uncle's call earlier that day. Indeed how her mother would benefit if any of Eryk's presentation were proved true. How was her mother now? Should Marzena have left the conference, paid the extra money for cancellation of her prearranged schedule and flown to Gdansk to see her? Maybe she should call again and talk directly with the hospital where her mother was or better yet with her mother to find out how she was actually feeling. Since it would be late night there, considering her mother's need for a good night's sleep, Marzena decided to wait for daybreak the following morning. She furiously checked the time lag all night. Finally she dozed off fitfully in early dawn, deciding to bring her mother here to the US as soon as she was strong enough to travel so she could be given the best cancer care here. If what Dr. Eryk Cyrek was talking about actually worked, then cancer patients would surely have alternatives to chemo and surgery.

Marzena was woken up almost instantly, or so it seemed to her, by her phone. Her uncle called saying that her mom had just passed away. Marzena chided herself for her erroneous decision a few hours ago. Her dad had not been wrong after all or devious. Had he really wanted her to be by her mom's side before she passed away? Sorrow for her mom, deep resentment for her dad, combined with guilt, at not being able to trust him and not being able to hold her mom's hand when she passed away, wracked her slim body with sobs. When she recovered, she reasoned that even if she had listened to her uncle or her dad and cancelled her previous plans, she wouldn't have been able to make travel changes from her conference in Seattle to reach her mom's bedside in Gdansk in time to actually bid her goodbye.

* * *

After more than seventy-two hours an exhausted Marzena arrived at Gdansk with a small carry-on and dark circles under her eyes. The plane had been delayed due to bad weather and last-minute travel plans had led to escalated ticket prices. With a sore mind and body Marzena went straight to her mother's funeral in the chapel by the hospital. Her uncle had arranged for her mother to be embalmed till she arrived. When they finally brought in her casket and Marzena reached out and touched her mother's hand, she couldn't help feeling an inner relief that this dear lady had not had to suffer long from her dreaded disease. But then if she had found out sooner about Eryk's research, her mom might still have been alive. For that she prayed for forgiveness from this special lady who had brought her into the world, seen her through her traumatic childhood, encouraged her in her youth and had not given way to self-pity when her marriage had broken up so suddenly. Riddled with questions and worries as she was, when she held her mother's hand, a strange sense of blessing seemed to emanate from the ornately decorated still form in front of her. It was as if her mother had come to life again. She was holding her daughter close kissing her forehead saying everything would be fine and that she shouldn't grieve for her mother's death but rather celebrate her life. Marzena closed her eyes and as tears rolled down her cheeks, she knew wherever she was, her mother would always be in her heart. It was an all-encompassing sense of affirmation of life not the negation of death. And then it started. The gnawing pain she had been feeling for the last few months in her lower abdomen rose in waves and became so excruciating that she doubled over and fell on the floor.

9

KAMOLA

The Big Yellow Moon

HER MOTHER'S SUDDEN PASSING made Kamola almost a grown up very quickly. Immediately after her mother passed away, when she was merely twelve years old, she would knock softly on her father's door with a "Baba, it is time to get up," very early in the morning. Her limbs were loose with sleep, her long hair tangled in knots. Her father scolded her for this. But Kamola knew somehow that it was her responsibility to take up some of her mother's domestic tasks. She cooked for her father and Shibu and Ramen, washed dishes, cleaned the house and kept their clothes washed. Besides doing all this household work, she started doing something else last year. It helped her deal with the loneliness she felt after her mother died.

Three Daughters, Three Journeys: Quest for Cancer Cure
Ananda M. Chakrabarty, Jill Charles, Indrani Mondal,
and Ranjita Chattopadhyay
Copyright © 2017 Pan Stanford Publishing Pte. Ltd.
ISBN 978-981-4745-90-1 (Hardcover), 978-1-315-19666-4 (eBook)
www.panstanford.com

Very recently some changes took place in her village. There was a new health center built by a group of rich foreign doctors in Palampur. Kamola came to know that their goal was to help the villagers to provide drugs and treatment for complex diseases which could not be treated in the existing local medical facilities. It took about two and half years to build the facility. It was a four-story brick building painted yellow with terra cotta borders. The building was surrounded by a grassy lawn and a garden. The garden had flowers that were native to the area. Along with some foreign doctors, Kamola saw several nurses and a few doctors with the same skin color as hers. What she understood from the conversation of the village elders that the new medical center was owned and run by a group of American and Indian doctors based in America. They needed local doctors, nurses and other staff members for its daily operations. Kamola herself had started working there since last year. Her job was helping in the kitchen. She had nothing to do with the patients. She served tea, drinks and food to other employees of the clinic. How she wished it was there when her mother was ailing! It was because of her mother's insistence that Kamola went to Palampur Primary School and finished fifth grade. She could read and write simple sentences in both her native Bengali language and English. That knowledge helped her to get the job.

Kamola thought about her mother's symptoms again. An idea flashed in her mind. Maybe that was what she would do. She would talk to one of the lady doctors at the clinic. For the last few weeks, she had been experiencing some discomfort at her left breast. It was even hurting a little from time to time. In the beginning, Kamola ignored it completely. Then one day, while she was drying herself after taking a shower, she felt something hard and lumpy at the right side of her left breast. She was so scared. She could not think about asking anyone what to do. At moments like this, she missed her mother the most. She did not have an older sister, an aunt or sister-in-law to talk to about this. She did not even have a girl friend in whom she could confide. But she saw a tinge of light in the darkness now. She trusted the doctors in the clinic. She had seen families leaving the health center happy and smiling. She was unable to comprehend fully what exactly went wrong with the patients who stayed there. But she could tell that the doctors and nurses were very devoted. In fact their sincerity inspired Kamola too. She was a loyal and trusted employee there. She did not let her personal emotions come in the way while she was at her job. She would ask one of the doctors in that clinic for help.

That evening Kamola went home a little sooner than usual. The house was empty. Kamola's father, Shibu and Ramen were enjoying themselves at the dance. She was happy for them. Her father tried his best to save his wife's life. But nothing worked. Even with his meager income of a farmer who worked on someone else's land, he did everything that the local physician asked him to do. He was also his father's childhood friend. Kamola accompanied her mother a few times too. As soon as they would walk in his clinic the doctor would ask her mother, "How is the pain today Boudi?"

Boudi, yes. That was what he used to call her. Her mom sighed. Kamola now realized that she probably did not want to talk much about the pain in front of her.

Kamola did not know what was happening to her that spring night. As the moon rose big and full, a lot of thoughts from four years ago were coming back to her. Kamola never saw her grandmother. She heard some stories about her from her mother. In course of her conversation with the physician, her mother once told him that she was certain that the abdominal pain would eventually kill her. The physician was surprised and asked his patient why she said so. Niru replied, "I heard that my mother suffered from it too. And finally the pain made her so exhausted that she did not have any energy left to breathe."

That night some unknown fear crept in Kamola's mind. She could not eat dinner. She made the plates ready for her father and brothers in case they came home hungry. As she drifted off to sleep she decided to do something about her worries and anxieties.

The next morning dawned pink and purple, the promise of a lovely spring day. The big Krishnachura tree that stood at the bend of the road leading to the new hospital building nodded in restless southern wind. The tree had burst into bright red blossoms. The cuckoo's call was loud and clear. Kamola woke up. She finished her morning chores quickly. Shibu, Ramen and their father came home late last night. But they had left already. They did not have a choice. If they chose to stay home and rest they would not get paid. Sometimes Kamola wished she had studied further. Then she could have had a decent job and would be able to help her father and brothers more. She had to stop because her family could not afford paying the school fees and buying books. Maybe a little more formal education would have helped her understand life a little better. Anyway, she did not waste much time pondering. She would have to act on what she decided.

10

SELENA

Knowing the Worst

When Dr. Weiss called Selena into her office to discuss the test results, Selena already knew they were positive. Her only question was whether she had the mutation on the BRCA1 or the BRCA2 gene or on both genes. Sometimes the test gave inconclusive results but Dr. Weiss would have told her that over the phone.

Dr. Weiss led Selena into her office, offered her a chair and shut the door. She sat next to her and together they looked over the paper with the diagnostic results.

"These are your test results, Selena. You tested positive for both mutations, on the BRCA1 and BRCA2 genes."

Three Daughters, Three Journeys: Quest for Cancer Cure
Ananda M. Chakrabarty, Jill Charles, Indrani Mondal,
and Ranjita Chattopadhyay
Copyright © 2017 Pan Stanford Publishing Pte. Ltd.
ISBN 978-981-4745-90-1 (Hardcover), 978-1-315-19666-4 (eBook)
www.panstanford.com

"I want the preventative surgeries," Selena said. "I want both the mastectomy and the oovectomy. The type of breast reconstruction I would want is…"

Dr. Weiss interrupted to ask, "Have you discussed this with your family?"

"It's my decision."

"Have you told your dad or any of your relatives that you're having genetic testing? I know you planned to let your aunts and cousins know your test results."

Dr. Weiss could perform the mastectomy and oovectomy, could refer Selena to a plastic surgeon, but the fear and worry Selena felt, and her family's concern for her could not be neatly removed or stitched up. With blue eyes full of concern, the doctor held Selena's hand and waited for her to speak.

"I'll tell Dad," Selena said. "I'll tell them all."

* * *

"Wake up, Princess, I'm here."

Slowly Selena opened her eyes and focused on her father, reaching out to him with her right hand, the one without the IV threaded into it.

The thin hospital blanket and gown covered her from the neck down, concealing all the bandages and scars. She glanced down at her chest, aching and itching under the compress, where her breast tissue had been removed and saline gel sacs had been inserted. Later, when her father had gone, she would run her hand over the temporary shape, careful not to disturb her bandages. Her chest hadn't felt this sore since puberty. Would she feel normal, weeks later, after the saline sacs were removed and her thigh fat was transplanted to rebuild her chest? At least it would be her own flesh, nothing plastic under her skin. She felt alien to herself now.

"How do you feel?" Dad asked.

"Okay, just a little sore."

"Do you need more pain medication?"

"No, Dad. I'm just trying to wake up now and it's not that bad."

"Have some water," he said, pouring her a glass from the plastic hospital pitcher.

The ice-cold water refreshed Selena's parched throat. She had been in surgery and recovery for hours.

"I brought you something," Dad said.

He reached into his backpack and brought out her iPod and earphones.

"So you can have music," he said. "Grandma Grace and I recorded ourselves singing in case you need a lullaby. Try track one."

"You recorded Grandma Grace?"

Unlike Selena's bluesman father, her Grandma Grace refused to perform. She sang at home and at church, for God and her family and always in a crowd of other voices, but her alto voice rang out beautifully over Dad's guitar notes when Selena slipped on the headphones.

"Keep on the sunny side, always on that sunny side. Keep on the sunny side of life. It will help us every day, it will brighten all the way if we keep on the sunny side of life."

More songs followed, blues and rock, Selena's favorites and more songs by Dad and Grandma: "When the Saints Go Marchin' In" and "Lonesome Valley."

"You've got to walk that lonesome valley

Well you gotta to go by yourself

Well there ain't nobody else gonna go there for you

You gotta go there by yourself."

Selena cried a few tears but she felt Dad and Grandma lifting her up with their voices, healing her more than the painkillers, even more than the protective surgery could. Selena wiped her eyes and smiled, looking up to see Grandma Grace standing in the doorway.

"Come in, Grandma," she said. "It's okay. I'm okay."

Grandma Grace's eyes were wet but she drew herself up straight and sat on the edge of Selena's bed to hug and kiss her.

"You're brave, Selena," she said. "You just keep on bein' brave."

Grandma Grace had been disappointed that Selena would have the oovectomy. She had asked "Couldn't you get married and have a baby first?"

Grandma Grace wanted their family to go on, but she never blamed Selena or called her selfish when she explained. "I don't want children like Claude does. I realize he wants to be a dad more than he wants to be with me. You have plenty of grandchildren and great-grandchildren, Grandma. This world isn't going to run out of Ramoses. And whether I marry or adopt, whether I ever have a child or not, I won't be alone, because of our family, because of you, Grandma."

"Are they taking good care of you, honey?" Grandma Grace asked.

"Yes. The doctors explained everything to me, about the surgery and the follow up. The nurses have been great, especially Rhonda."

Grandma Grace gently washed Selena's face, then brushed and braided her hair.

"We had to take Jojo over to our house," Dad said. "She keeps sniffing around, looking for you."

"We should bring her in here," Grandma Grace said. "We just need Dr. Weiss to write a letter saying that she's one of those 'emotional support animals'."

Selena laughed.

Grandma Grace straightened up everything in the already neat hospital room and wanted to feed Selena. Dad and Selena insisted that her stomach wasn't steady enough after surgery, but she left her a banana and homemade corn muffin for the next day "when you start to get hungry."

"We'll be back tomorrow at eleven to bring you home," Dad said. "I wish we could bring you home now."

"Don't worry, Daddy. The doctors just need to make sure I'm healing up and it's just for one night."

Grandma Grace spread Selena's bright patchwork quilt over her gray hospital bedspread before they hugged and kissed her goodbye.

Selena felt lucky to be so loved, listening to her music under her quilt and reading *A Day Late and A Dollar Short* by Terry McMillan. McMillan's novels about black women's friendships and love troubles always took Selena away from her own problems, and yet reminded her of the good folks in her life. She texted her friends to let them know "Surgery went OK. Resting up and going home tomorrow. Thanks for all the love and prayers."

Janine had helped her best friend check into the hospital before her surgery and left her with a new graphic novel, *Weird Girls*, about two high school nerds based on Janine and Selena. On her phone Selena checked Facebook to see two dozen students from the high school holding up a pink sign with rainbow letters reading "We love you, Ms. Ramos! Live strong!"

She would go back to school tomorrow, back to work the next day. Her breasts would be rebuilt with her own thigh fat and her own nipples. No one who looked at her fully dressed would ever know she'd had a mastectomy. The scars would be minimal. She worried more about the oovectomy, not knowing how soon menopause might follow and worried about feeling hot, tired and moody when it reached her. It took months just to schedule the surgeries, to make sure no infection followed each one.

In spite of the surgeries and the genetic defects, Selena knew her body was healthy enough to bicycle ten miles, to lift 50 pounds easily, to climb three flights of stairs without gasping. She would live and thrive for decades, unafraid of the shadow of cancer. No tiny evil cell could steal her life from her.

In the evening, Rhonda the nurse moved another patient into the next bed in the hospital room and politely introduced her and Selena to one another. Rhonda drew back the gray curtain around the second bed as a Latina lady in her fifties wheeled herself in with a wheelchair. The other patient wanted to walk around, Selena could tell, but she listened to Rhonda and settled into bed. Selena noticed her wooden bracelet showing the Virgin of Guadalupe.

"Teresa Avila, this is your roommate for the next sixteen hours, Selena Ramos. You are both great ladies, both recovering from a double mastectomy and I bet you'll find other things you have in common. I'll be back to check on you at nine and if you need anything before then, just call."

"Pleased to meet you," said Selena.

Teresa said the same in a soft Spanish accent. Teresa had kind eyes and Selena felt at ease with her, in spite of, or maybe because of, their shared pain.

Teresa glanced at Selena's bandaged chest. Her own white bandage showed in the V-neck of her hospital gown.

"You're very young to have cancer," Teresa said.

"I haven't had cancer yet," said Selena. "I got the mastectomy to prevent it. I have the BRCA1 and 2 mutations and my mother died of ovarian cancer when I was eighteen. Four of my aunts had cancer and my grandma also died of it. I decided to have my breasts and my ovaries removed to prevent it."

Teresa's eyes widened as she heard Selena's story, then she nodded calmly.

"It's good you can prevent it, Selena. Your family shouldn't have to lose anyone else. I've had breast cancer twice. The chemo left me so weak I couldn't lift my head off the pillow. My husband had to carry me."

Teresa looked short and light enough to carry.

"How long ago were you diagnosed?"

"The first time was when I was fifty-one. I just had one tumor so I had a lumpectomy, then chemo and then radiation. I took hormone blockers for five years after that. When I had cancer I was working as a paralegal at a big law firm downtown. They did all debt collection cases. The lawyers were so busy I don't think they noticed when I left work an hour early for chemo, not even when I replaced my hair with a wig. It *was* human hair in exactly my same color and style. I didn't want them to know I was sick."

"Why not?"

"Honey, they would have fired me. I would do my monthly chemo in the afternoons and tell them I was dropping off documents at the courthouse. They never knew and they never asked me anything about my life."

Selena felt grateful for her fellow students and teachers at the University of Chicago and the high school, for the women at the shelter.

"Weren't you tired the next morning?"

"I was tired all that year. Paolo would make me yerba mate tea and a big smoothie of mango and banana and vegetables. Paolo, my husband, is from Paraguay and I'm Mexican, born and raised in Chicago. Paolo would drive me into work and pick me up every day."

"He sounds like a great guy," Selena said. "And you look younger than fifty-one."

Teresa chuckled and said, "I'm sixty-seven now."

"Do you still work at the law firm?" Selena asked.

"No. I had to quit work this time. They had been pushing me to retire for years. I only told two of my friends at work about my cancer and they kept it quiet because I needed the work and the insurance. I'm so worried about the bills from this. Paolo is retired too and our children are helping us."

"That's good. Do your children live in Chicago?"

"Yes. Felipe and Cecilia both live here, with their spouses and kids."

Teresa dug through a substantial brown leather purse and found her purple cell phone. She showed Selena a family picture.

"There's Cecilia and her husband Marc. Their kids are Natan and Christina. And that's Felipe and his wife Eliana and their girls, Gloria and Mariana."

"You have a nice big family," Selena said. "I'm single but I have a lot of family here too."

She wasn't about to bring up Claude and whether they were a couple or not.

"It was so hard telling my children I was sick again. I'd been healthy for so many years. I had some arthritis from the chemo but that's normal. It kills some healthy cells along with the sick ones."

"I hope you have a full recovery," Selena said. "If it's all right with you, I'll pray for you."

"Thank you," said Teresa. "All good prayers are all right by me and I'll say one for you too. I'm in stage III, you know, but I never give up."

Selena wished she could do a lot more to help Teresa. They exchanged phone numbers and emails, deciding to meet again after they left the hospital.

"I go to a cancer survivors' group," Teresa said. "They also counsel family members who have lost someone to cancer. You could go too if you want, to talk to others who have lived through the same things."

"Thank you," said Selena. "I feel like I've had it so much easier than so many women with cancer. What could I teach the survivors?"

"Having it young, even having the surgery you had, is hard in a different way. You are a survivor too. You have been since you lost your mother. The group would accept you. We accept all kinds, young, old, all races, gay, straight, even two men, one with prostate cancer and a man with breast cancer, an old Marine officer exposed to chemicals in the Iraq War. They are the kindest people but we never would have met otherwise."

"I'm glad I met you, Teresa."

"I'm glad I met you too, Selena."

Despite the constant bright lights and footsteps in the hospital hall, both women fell into a sound sleep in their shared room.

11

MARZENA

Night

THE NEXT THING MARZENA saw when she opened her eyes was like a dream. She was staring into bright blue eyes. The eyes looked very familiar but turning away she realized that the stark white and sterile surroundings around her were not. Her head felt fuzzy. Was she in a hospital bed?

She quickly brought her eyes back to what was in front and found herself looking again straight into the blue eyes of Dr. Eryk Cyrek. "How on earth," she began. "Why am I here? What happened?" Incoherent questions rushed out of her in a stream.

Eryk pulled up a chair and sat by her bed. He took her arm and checked her pulse. He gave her a half-smile making her heart skip a beat. "Actually I'm here

Three Daughters, Three Journeys: Quest for Cancer Cure
Ananda M. Chakrabarty, Jill Charles, Indrani Mondal,
and Ranjita Chattopadhyay
Copyright © 2017 Pan Stanford Publishing Pte. Ltd.
ISBN 978-981-4745-90-1 (Hardcover), 978-1-315-19666-4 (eBook)
www.panstanford.com

to present my paper on 'Alternate Remedies for Breast and Ovarian Cancer' and so working with the oncology department of this hospital. They have allowed me to look into your mother's medical reports for case history. I was at her funeral in the chapel and you passed out holding your mother's hands." He looked concerned. "Tell me has this happened before?"

Marzena felt dewy-eyed with all this attention but her pain was still quite intense. She gestured to her abdomen and said, "I've been having these aches and pains lately but none as terrible as this one. I think I'm just tired after my long flight. But I have go back to the funeral," she ended with more vigor than she felt.

"Not really, Marzena, I believe that's your name, and I just found out we are alumni from the same high school here in Gdansk. So now I really have to get you up and going. Everything is being managed quite well at the funeral by your dad and uncle. I think you just need to relax here while we run some tests," said Eryk squeezing her hand briefly before turning away.

"I can stay here forever looking up at you" was on the tip of Marzena's tongue but then another thought struck her. "Hey Eryk, weren't you at the Needle Phobia Conference in Seattle last week? I was presenting a paper there and you commented on it..."

"Yes, of course. That's just what I was telling you. I think everywhere I go I keep trying to make people aware of alternative cancer cures, at all the presentations I attend..." Eryk's voice trailed away as another bout of pain knocked the breath out of Marzena.

The next few hours passed in a numb whirl for Marzena. More pain, less pain, more consciousness, less consciousness whizzed by...When she finally woke up she found her uncle Patrick and her dad sitting by her bedside. They looked grave but relieved to see her eyes open.

Marzena had last seen her father when she had gone to study nursing in the US. It was a long time ago but it seemed like yesterday. Except for a slight graying at the temples he was the same picture of strength and care. What had happened, she wondered, that he had to leave her mother when she'd needed him most?

She felt tears rolling down her cheeks and as she looked away trying to control herself, her dad reached out and took her hand. "This is not the time to look back, my dear. Do you remember Eryk from your high school? Well! He's doing some groundbreaking research and collaborating with our hospital for cancer case studies. So there is no need for you to lose hope at all." He held her hand tight in a warm comforting clasp.

"It's too late now anyway," Marzena heard herself say in a small voice as she wriggled her fingers out from her dad's grasp. "I wish I'd known of all this new research when mom was still alive. Her doctors could have used these treatments to her best advantage then."

Her dad smiled brightly, "We can't give up hope. It's not too late. Maybe they'll find spectacular symptom-arresting results in you..." He bit his lip and stopped short.

Marzena's uncle leaned over and patting her shoulder, said reassuringly, "You can stay here for now till you're all better. You know you don't have to worry about anything, right?"

Marzena looked up at both of them blankly. What was going on? Why couldn't she move in her bed? She was dazed and trying to look at her surroundings again, she moved her head to the side. Horror of horrors! She was stunned to see a needle in her left arm pumping some drugs into her body. No wonder she couldn't move. She lost control. Her old needle phobia that she had been trying to overcome as a nurse and had fought furiously against when she had seen her ailing mother in chemo with needles around her, came crashing back. She shouted with anguish and anger. "Get this...this evil thing out of me at once!" she yelled hysterically. Her body trembled convulsively as she arched over suddenly and tried to pull the needle out with her free hand.

The emergency patient bell beside her bed rang out sharply and two attending nurses rushed in and along with her Dad tried to calm her down. But she struggled furiously and in panic jumped out of the bed and started running to the door. She didn't go far though. Marzena ran into a rather warm obstruction that held her strong and without hurting her, she was quickly brought back to the bed.

It was Eryk. He smiled soothingly at Marzena as he gestured to the baffled attending nurses to tuck Marzena in. Marzena felt a strange calm enveloping her as she looked up at Eryk, exuding health and hope. Shaking hands quickly with her disheveled dad and upset uncle, he turned to Marzena. "Your family thought I should be the one to explain all your options before your doctor talks to you."

As he again pulled up a chair near her bed she couldn't help thinking rebelliously why weren't they letting her get up and leave?

"If you remember anything from what I was trying to explain at the Seattle conference," began Eryk, "I'm collaborating with a team of medical scientists and have come up with an advanced and patient-friendly version of the anti-cancer and cancer preventive protein called Neelazin. Right now this can be administered as an intravenous injection but I've been researching diligently to make it equally effective orally."

Eryk brought his chair even closer to Marzena's bed. She looked from one face to the next puzzled and uncomprehending.

Then her father took her hand, "My dear child, you have been diagnosed with ovarian cancer. But I have already talked to the oncologist about your needle phobia. So you could perhaps do trials for some of Eryk's new cancer cure breakthrough!"

Eryk explained to her carefully that while Marzena was sedated they had taken an MRI image of her abdomen and found a tumor in one of her ovaries. Because her mother had succumbed to breast cancer, further tests revealed that Marzena had inherited a kind of mutation called BRCA1 or BRCA2 that made her vulnerable to breast and ovarian cancer. Subsequent biopsy of her tumor led to the diagnosis that it was malignant ...Eryk went on talking.

Marzena's head was spinning. She couldn't be hearing right! A budding ovarian cancer in her body? No, that couldn't be true! She was just twenty-two for heaven's sake! She couldn't die, she had so much to live for! She had to get back to her promising nursing career soon! She dropped her dad's hand, reached out blindly, caught Eryk's hand and half-sitting up on her elbow, she cried out passionately, "I have to get better, Eryk. I love my job as a nurse. I have to go back to all those children in need for help. They mean the world to me! I can't let them down!"

She gulped back tears and said in a broken voice, "Just get these damn needles out of my body and remove all my ovaries once and for all, so I can be all right again," she pleaded vehemently.

The room fell silent after Marzena's fervent outburst. The three people by Marzena's bedside looked at each other in silence. Then Marzena's dad broke the silence, "I do understand about your needle phobia, my child, and you're too young to get rid of your ovaries. As your dad and above all as a doctor whose main concern is to cure his patients...I can't agree to that...you have your whole life ahead of you...besides since you have tested positive for the BRCA mutations you have the risk of the cancer showing up elsewhere like your mother ..." His voice trailed away.

All four stared ahead as if peering into the future. Although the room was brightly lit, it seemed from the look on their faces that they were finding it hard to see.

Eryk stood up and started pacing up and down the room. "In my research I have used some of the same techniques of chemical modification on Neelazin that the insulin people are using to make oral insulin. As I have said repeatedly, it works!" Eryk stopped in front of Marzena's dad and brought his closed right fist down hard on his left hand with vibrant hope. "I have tested my oral Neelazin tablets dissolved in water not only in tumor-bearing mice but also on the model called the onco-pig model. Believe me it not only works well with these higher animals like pigs and monkeys but we have also proven it to be without any toxicity whatsoever."

Eryk started walking again and this time came and stood by Marzena's bed. Looking down at her with his bright blue eyes he continued, "Part of the reason I'm here in Gdansk is because I got authorization from regulatory agencies and the hospital ethics committee and my research supervisors to use this drug in a clinical trial on cancer patients here in this main hospital. As I told you I was planning to enroll your mother also in my large clinical trial."

He stopped, bent down and took Marzena's hand almost as if he was proposing to her, thought Marzena with a wry smile. "Will you give me your consent to be part of this phenomenal trial? Remember there will be no needles, just oral tablets to take. It may take several months though and you will have to get leave of absence from your job," he ended in a sober voice but with expectant eyes.

As Marzena gazed up at Eryk, her mind flew back to her high school days. Was there a reason to the strange attraction she had felt for this handsome young man even then? Was it because one day he would be able to offer her the key to life and death as he was doing now?

Eryk found he was holding not only Marzena's hand but also his own breath as he waited for Marzena to answer. He saw before him a fragile form with a strong spirit and hoped with all his might that this young girl, alumni from his own high school, as he had recently found out, would be brave enough to try out this experimental drug, not only for her own sake but for posterity. He smiled encouragingly at her and squeezed her hand.

Marzena broke into his thoughts in a low voice but with deep conviction. "Eryk, I trust you completely," she began simply. "If it helps you and me to take this experimental oral Neelazin tablet to heal my cancer, so be it! I think I'm more excited than you! As your patient and trial subject, I'll get the unique chance to show the success of your newly minted drug. Just think how much hope and benefit it may bring to so many cancer patients across the globe. And," she gulped, "if things go wrong, it was all for a noble cause…" Tears trickled down her cheeks. They were as much tears of untold sorrow as of unspoken hope.

12

KAMOLA

The Milk Tea and Sterile Bed

KAMOLA WAS READY TO go to the hospital. Usually it took her about twenty minutes to get there. For some reason it took her longer that day. Or at least it seemed longer. Once inside the building she went straight to the kitchen. The very first thing she did was put the saucepan on the stove and pour milk into it. Voices from the next room came to her. She knew before long three of them would saunter into the kitchen in search of their first cup of strong, milky tea. They were all doctors, two males and one female. Kamola could not make out the names of the male doctors. Neither of them spoke the language of Palampur, which was Bengali. They were white people. Kamola could not understand a single thing they said. They took their tea without any sugar. The lady was

Three Daughters, Three Journeys: Quest for Cancer Cure
Ananda M. Chakrabarty, Jill Charles, Indrani Mondal,
and Ranjita Chattopadhyay
Copyright © 2017 Pan Stanford Publishing Pte. Ltd.
ISBN 978-981-4745-90-1 (Hardcover), 978-1-315-19666-4 (eBook)
www.panstanford.com

young and quite attractive. She had the same skin tone as Kamola's. She could speak with her foreigner colleagues in their language. At the same time she was perfectly capable of communicating with the patients and other employees of the hospital in their native language. She was always busy checking on the patients, listening to their discomfort, instructing the nurses or reading big books in her own office. She worked hard. Some days she would forget to come to the kitchen to get her lunch. But if she had time she laughed, cracked jokes with her colleagues. One day Samir came looking for Kamola. The lady doctor was there in the kitchen at that time. She teased Kamola about him. All the patients loved her charming personality. Kamola made up her mind. She would talk to her.

There was a moment of hesitation. Kamola never addressed her directly. After all she was a mere kitchen helper in that modern hospital. She heard other patients calling her Dr. Mishra. Dr. Rini Mishra was her full name. The morning air smelled of boiled milk, dish detergent and Rini's body lotion.

Kamola said, "Doctor didi, I want to talk to you."

"About what?"

There was a momentary silence.

"About myself."

Rini signaled to her two other colleagues to go ahead in their morning rounds without her.

"What happened, Kamola? You look worried."

Tears rolled down from Kamola's eyes.

"I don't know. It is just that…"

"Just what, Kamola?"

"I have been having some pain in my left breast off and on."

"For how long?"

"About seven or eight weeks."

"Why don't you come with me right now? I will examine you. Is that all right with you?" As she was saying those words, Dr. Mishra motioned Kamola in the direction of Examination Room 10. It was located all the way at the end of the hallway to the right of the kitchen. Kamola followed her silently. The room was tiny. The walls inside the room were painted light blue. In all those days of working in the clinic, Kamola had never stepped inside one of those rooms. She lay on the bed as Dr. Mishra instructed her. She asked Kamola to undo her cotton blouse. Kamola was not sure for a brief moment. But then she did what she was asked to do. From that moment onwards the new journey for Kamola began.

13

SELENA

No Hard Feelings

SELENA MET CLAUDE AT a small dark wood bar downtown, a place she never liked except for its French 75 cocktails. Rain tapped on the windows as Selena settled into a quiet corner booth with her usual drink.

Claude's calls and visits had become rarer and rarer for the last five months. They no longer talked about living together, engagement or their future, yet neither talked about leaving the other either. Claude told her he feared hospitals too much to visit her there after her mastectomy or her reconstructive surgery. Now three days before her oovectomy, Selena had started to accept that she might never hear from Claude again, that he had given up on her after three years in love, when he called her one day and asked to meet after work.

Three Daughters, Three Journeys: Quest for Cancer Cure
Ananda M. Chakrabarty, Jill Charles, Indrani Mondal,
and Ranjita Chattopadhyay
Copyright © 2017 Pan Stanford Publishing Pte. Ltd.
ISBN 978-981-4745-90-1 (Hardcover), 978-1-315-19666-4 (eBook)
www.panstanford.com

Waiting for Claude, she glanced around the familiar bar with its dim lights and muted TV sets showing Chicago's sports teams playing basketball, hockey and football. She'd waited here so many evenings while Claude finished up work on a computer program, while he texted her again and again to apologize for being late. *I'm never coming back here*, Selena thought.

Just then the door opened and Claude came in, wrapped in a soaking wet black raincoat. He hung it from the coat hook on Selena's booth and sat down across from her.

"Sorry I'm late," he said.

He ordered a beer from the waitress and Selena resisted ordering one more French 75 cocktail. The lemon had sunk low in the ice of her drink, like the sun on a winter evening.

Claude and Selena gazed at each other, neither one wanting to speak first. They reached across the table and held hands.

"I love you, Selena."

"I love you too, Claude."

"You don't have to do this," he said.

"Yes I do."

She had told Claude her test results and her plan for preventative surgeries even before she told Daddy or Grandma Grace. Grandma Grace had burst into tears. Each of Selena's cousins listened patiently to her test results, hugged her and told her it was her decision. Her cousin Marta thanked her for getting tested.

"I'm getting tested too. I don't care what Joe thinks, I want to know. I need to know what my cancer risk is and I'm done having kids."

Daddy hadn't tried to talk her out of the oovectomy. He only said, "Whatever you need, Princess. It's up to you whether you have children or not."

Grandma Grace asked Selena to consider waiting on the oovectomy until she married and had at least one child. "You'll regret it when you're old, Selena. Your mama was glad to have you, right until the end."

"I know I don't want to pass the mutant gene on," Selena said. "And I'm really questioning if I ever want to have any children, now that I'm grown up and working and see how hard it would be. Claude has never even baby-sat, yet he's determined to be a father. He wants that more than he wants me."

"Maybe Claude's not the one for you, and he's not the only one you can ever love," Grandma Grace had said. "Don't give up on it."

"I won't. The oovectomy doesn't mean giving up on love. It means fighting cancer. I can't risk dying at forty because of a man I might someday meet or a child I might someday want. If I want one that bad, I'll adopt and anyone who loves me will have to marry me as I am."

In the dark bar Selena and Claude felt isolated, even from each other. The bartender and waitress barely noticed them. The last few drinkers walked out into the rain.

"I can't marry you, Claude," Selena said.

He looked down but didn't let go of her hand.

"Would you feel the same if your test results had come back negative?"

She nodded. "We would have found out sooner or later that it won't work and I'd rather know that now than two years from now, after we'd married and I was home alone with the baby while you were at work programming for sixteen hours a day and we still weren't saving any money. I can't parent for you and I can't pretend to be anything I'm not."

"You wouldn't be alone with a baby. That's not loneliness. You've never been alone, Selena. You weren't alone before you met me. You had your folks, Grandma Grace, Janine, Marta, Clarice and all the cousins. I'll bet you find another boyfriend in a month without even trying. You'll hardly notice I'm gone."

"I notice now."

She blinked back tears.

"Then don't leave me. Some women wait years to find someone to marry and have children with."

"I need to be me, more than I need to be a mother. I can do more good as a psychologist and I am happy without kids. But I can't ask you to give up your dreams for me. You need to marry someone who wants children as much as you do, Claude."

"I couldn't ask you to risk your life for my dreams either," Claude said. "I do love you, Selena. It hurts so much to think of letting you go."

She squeezed his hand and said, "I want you to have the best possible life."

"I want the same for you," he said. "We had three good years and none of that time was wasted. I still want to see you sometimes, to know what's going on with you."

"Come visit me in the hospital after my oovectomy, just like my friends."

"I can never be just your friend." Claude admitted.

He leaned forward and kissed her softly on the lips. They stood up and hugged each other for a long time.

Claude insisted on paying for their drinks, as always. The rain had slowed down and Claude walked Selena to her car, then stood under a golden streetlight watching her drive away. She looked back in the rearview mirror, again and again, until she turned toward Lake Shore Drive to rush home between the lighted city skyscrapers and the endless black of Lake Michigan.

Selena knew Claude would never visit her in the hospital. She didn't blame him or resent him for that, but she knew. She would not call him, not email him, not text him or even look him up online until many years later. It tore her heart to say goodbye to a kind man who still loved her, but it would have hurt them both even more to stretch the love between them like string until it unraveled and finally broke.

14

MARZENA

New Dawn

THE TREATMENT BEGAN AND Marzena was given two tablets of oral Neelazin with a glass of water every day, one in the morning, one at night. Marzena's tumor was closely monitored. Her dad had insisted she stay with him at his new home with his new wife during this stressful trial period. But he insisted on calling it her recovery. Her stepmom was courteous to her and along with her dad really tried to be kind without being intrusive. After a few weeks with them, Marzena moved back to her old family home where her mom had lived alone till she had passed away. The wooden beams in her bedroom still had her initials etched in them from years gone by and she couldn't hold back an errant sigh

Three Daughters, Three Journeys: Quest for Cancer Cure
Ananda M. Chakrabarty, Jill Charles, Indrani Mondal,
and Ranjita Chattopadhyay
Copyright © 2017 Pan Stanford Publishing Pte. Ltd.
ISBN 978-981-4745-90-1 (Hardcover), 978-1-315-19666-4 (eBook)
www.panstanford.com

thinking of those comforting family times with both her parents.

After two long nail-biting months, Marzena's tumor subsided and she tested cancer free. It was time for celebration but Marzena didn't want anything too fancy. Though hesitant at first, her dad agreed to her idea of a simple family get together with Eryk of course, in her old family home. Her stepmom proved to be a great cook and brought over a delicious home cooked dinner in honor of a brand new fully recovered stepdaughter. Eryk brought some wine and her uncle got her favorite glazed kolachke with raspberry and pine nut topping for dessert. It was a wonderful evening full of good old family banter and reminiscences. Marzena couldn't help bringing up her high school days and as Eryk helped her around the house with solicitous care she kept watching him from under her lashes to try to figure out if his feelings for her had taken on a deeper meaning as hers surely had. After her guests left she wandered around the house and looked at all her mother's needlework still hanging on their living room walls. She didn't know whether to laugh or cry hanging on to her dear mom's memory. Finally, after a week of some serious soul searching, she sold the property.

By that time Eryk had given her the green signal that she was cancer free and so free to return to New York. But as a preventive, Eryk prescribed her one tablet of oral Neelazin daily for the next six months.

After more than six months Marzena returned to work with Dr. Kim Wong. The whole department was thrilled to see her back and fully recovered. They brought her flowers and hugged her in welcome. As Marzena looked in on her new patients in their wards she realized it was not just a welcome back to her job but to life itself. She remembered how very depressed she had been when she had first found out about her life-threatening disease. But she knew now that she wouldn't have changed her destiny for the world. Her experiences had taught her how precious life was and had made her love her profession even more. She had learned to be a kinder nurse and a more generous human being.

As for Eryk, she was glad to have gotten the rare privilege to bring Eryk's dream to fruition, the same Eryk who had stirred such riotous color in her dreams so early in her life. It was as if she and Eryk were meant to be together. Granted, he had never really expressed his personal feelings for her but then she was his patient after all. She could never ignore his stress not only to prove his research results but also to get her better fast. She was just sure he knew that for her he was not only the reason she was able to look at the sunrise with thankfulness every day, but he was also her true love whom she would always want by her side to watch it with.

After Marzena was declared fit to travel back to the USA, Eryk had gone to attend an International Cancer Conference that he had been unable to attend for the last year as he had been busy with his research. This time he had been invited to present his highly applauded paper on "Oral Neelazin" along with clinical trial case studies, to endorse his research. He had promised to keep in touch and check on Marzena's progress. Marzena was sure it was a matter of time before he would come visit her in New York. She was convinced that he was already more than aware she had feelings for him for she had never tried to hide them anyway in the last few months in Gdansk. The next time they would meet, Marzena surely would tell him all about it in so many words.

Marzena's heart was full. If she missed anyone it was her mom but she knew wherever she was her mother would surely see how blessed her daughter felt every single day. She would be happy that Marzena had found a life to love and the love of her life.

15

KAMOLA

Light at the End of the Tunnel

DR. RINI MISHRA, A bright, young oncologist who was born and raised in Kolkata, India, to emigrant parents, immediately felt a pang in her heart as she examined the teenager. Rini's ancestors came to Kolkata from Rajasthan, the dry desert land of the western part of India, to earn a livelihood. They never went back. Eventually they adapted to the Bengali ways of life. Rini's Bengali was smooth and fluent, without any foreign accent. She went to study abroad after completion of her MBBS degree from the Calcutta Medical College. As she grew up, her grandfather always told her stories from the myths and legends of India. Maybe that was why the land fascinated her. Once she finished her

Three Daughters, Three Journeys: Quest for Cancer Cure
Ananda M. Chakrabarty, Jill Charles, Indrani Mondal,
and Ranjita Chattopadhyay
Copyright © 2017 Pan Stanford Publishing Pte. Ltd.
ISBN 978-981-4745-90-1 (Hardcover), 978-1-315-19666-4 (eBook)
www.panstanford.com

studies in the US, she came back to serve the people of her homeland. There was also another reason for her return. She did not want to think about it then. It was a personal reason.

Her experienced hands immediately detected the tiny lump in Kamola's breast. She pressed it gently.

"Does it hurt, Kamola?"

"A little."

"Do you see any secretion coming out of it?"

"I don't remember seeing any."

"Kamola, who else is there in your family? Do you have a mother and a sister?"

"No. My mother passed away four years ago. I never had any sister."

"How did your mother die?"

At this Kamola almost lost her composure.

"No one knows."

"What do you mean?"

Kamola described the terrible pain in her mother's belly and Dr. Mishra's face grew very serious. She wanted to know the name of the physician who saw her mother, prescribed MRI and a full body CT scan and biopsy of the lump for Kamola and asked her to bring her father to the hospital as soon as she got the reports of those examinations. She explained to her patient in very simple language what she needed to do next.

Kamola was thoughtful too as she left the examination room. By then she could figure out that there was a possible relationship between her mother's fatal illness and her own symptoms. What could it be? She took the paper that Dr. Mishra gave her and showed the lady in the testing room. She went through the experience of being under a CT scan machine for the first time. She was very scared but at the same time she had hope. She trusted Dr. Mishra. She felt she was going in the right direction. She closed her eyes tight as the biopsy needle was injected right where her lump was. She was trying to remember her mother's advice, "Kamli, do not forget, the night is the darkest before dawn."

Kamola did not come to work in the kitchen after the day of all those medical tests. After three days she came to the hospital with her father. In those three days she tried to tell her family as much as she could about her symptoms. It was difficult. In the environment where she lived, girls, particularly teenage girls like her, were not supposed to talk about their body parts to the men. Kamola had no choice. Even from the little she could communicate with them she could tell that she made all of them nervous. After her mother's death they started depending on her in a way that they themselves were not quite aware of. Now the news of Kamola's not feeling well and having those tests recommended by a doctor from the new hospital left them worried.

Kamola's father, a man of about fifty years, walked with a heavy heart to Dr. Mishra's office with his sixteen-year-old girl. He had aged more since his beloved wife passed away. The thought of sheer helplessness that he felt with Niru's pain still made him restless. He was waiting anxiously to hear what Dr. Mishra had to say about Kamola's reports.

Rini saw those results the night before. The biopsy report told her that Kamola's tumor was malignant. Rini thought for a long time. She could not imagine Kamola, a lively, sixteen-year-old girl going through not only lumpectomy, but also radiation and chemotherapy, the most common treatment for this. She would lose all that beautiful curly hair. Just the idea of Kamola going completely bald almost brought tears into her eyes. In her mid-thirties, she knew that it was not okay for her to be overly emotional with her patients. But she could not help it. The chemo would affect her healthy cells also, giving rise to diarrhea, nausea, vomiting and other symptoms. Life would be so different for her. She thought about the option of surgery including radical mastectomy. But what about her future? She was only sixteen. She had her whole life before her. Any of those options would ruin her chance of getting married, being a wife and a mother. She knew that Kamola had feelings for that handsome young guy named Samir. At this Rini somehow felt bonded with Kamola. How did it feel to be so young, innocent and in love? She was once there too. Was it all over or did Rini still have some sweet feelings for Eryk?

The memory of Eryk immediately turned her train of thought. A few years ago she did a residency in Midwestern Clinic in the central US with Eryk. They grew fond of each other, although nothing serious happened. Rini recalled with pleasure the reason she chose oncology as one of her specialties, not so much because Eryk also chose the same subject but because of an intriguing meeting she and Eryk had with a seminar speaker named Dr. Sam Roy.

Rini smiled because she and Sam both spoke their native language and she was fascinated by the drug that Dr. Roy developed, apparently from some kind of a marine bacterium from a remote Arctic island north of Greenland and Iceland named Blue Frost Land or something like that. Sam Roy discussed at length the clinical trials he conducted in a large number of cancer patients with a variety of cancers, mostly stage III and IV cancer patients with metastatic, drug-resistant tumors. Being very different from the rationally-designed chemotherapeutic drugs used in the treatment of cancer patients, the protein drug from the marine bacterium, which he named Neelazin, had unique efficacy against these drug-resistant tumors in stage III and stage IV cancer patients, allowing in most cases both partial and complete regression of the tumors, thus prolonging the lives of many of his patients. He showed in laboratory experiments that Neelazin had not only entry specificity in cancer cells, entering mostly in cancer cells but not in normal cells, but also cancer preventive activity. More importantly, as Dr. Sam Roy stressed, unlike the chemotherapeutic drugs, Neelazin showed very little toxicity and side effects in the patients. As a protein, even though it was small in size, just a tad bigger than insulin, Neelazin, however, had to be given by intravenous injections.

Rini vividly recalled, even today, the dinner conversations with Dr. Sam Roy. Both she and Eryk felt that it's easier for patients if they do not have to go through intravenous injections for weeks or even months if they could have a drug that can be taken orally. This is the topic that dominated their dinner conversations with Sam Roy. Dr. Roy was not convinced that a protein, even a small protein like Neelazin, could be effective if given orally because it could not pass through the gut intestinal membranes to enter the bloodstream to reach the tumors. Looking back at that evening and the dinner conversations, Rini also recalled the ideas that crept in her and Eryk's minds.

Could they develop an oral version of Neelazin, since Neelazin seemed so different in both being nontoxic and effective against drug-resistant tumors? While Eryk said that he would devote most of his time to developing an oral variety that could be taken as a tablet with a glass of water, Rini thought that the drug would still be expensive because of the protein isolation and bulk manufacturing costs that people in her native India and many impoverished countries would not be able to afford.

She read in the scientific literature that one avenue for bulk manufacturing drugs, perhaps even including protein drugs, is to produce them as part of plants, particularly common food plants such as tomato, lettuce, etc. That thought never left her mind and indeed soon after she returned to India, she started a collaboration with a friend at Jawaharlal Nehru University in New Delhi, a well-known plant biotechnologist, to express the Neelazin gene in tomatoes. Her friend, though initially skeptical and somewhat reluctant, understood the importance of the challenge and finally agreed. In a couple of years, her friend Dr. Datta, using the codon-optimized Neelazin gene, expressed the gene in tomatoes and demonstrated the strong anticancer activity in the Neelazin-expressing tomato extracts, but not in the native non-Neelazin-expressing tomato extracts. Dr. Datta's group further showed that when fed orally in tumor-bearing mice harboring human tumors such as breast, ovary, lung, colon and prostate, the tomato extracts could allow significant regression of the tumors in just six to eight weeks. Dr. Datta and Rini got excited. Could it work for higher animals and perhaps even in cancer patients? They gathered all their data on the histological analyses of no changes in the normal animal tissue but the total regression of the tumors in animals and presented the data to the Government Regulatory Agency as well as their Institutional Regulatory and Ethics Committees for permission to conduct clinical trials in patients with breast, lung, prostate, cervical, ovarian and other cancers. The approval for such a clinical trial came about only recently and Rini thought that perhaps Kamola could be included in this trial. She also decided to check if Kamola might have inherited the BRCA1 or BRCA2 mutation from her parents since the way Kamola described her mother's illness, seemed like it was a case of ovarian cancer, perhaps as a result of the BRCA type of mutation. She needed to talk to the physician who saw Kamola's mother to learn more.

In the morning she entered her office happy and cheery. She could see the light at the end of the tunnel. The treatment plan by including her in the clinical trial using genetically-altered Neelazin-expressing tomatoes was not only free during the trial but would be inexpensive and affordable by a low income family anywhere in the world. She saw the worried look in the middle-aged man's eyes. Somehow the father and daughter reminded her of her own childhood. She was happy the she could give them some good news about Kamola's future treatment plan if they agreed to include her in the free clinical trial. The cost of the Neelazin and the hospital stay of very sick patients would be borne by the foreign promoters of the Palampur hospital. She also knew that none of them had any knowledge to understand any complex medical terms. They would not understand if she started talking about the details of the inheritance problem of the genetic mutations, assuming that's what was responsible for Kamola's and perhaps some other patients' cases. She would find out soon by sending the biopsy sample to a local diagnostic company.

"Kamola, I saw your reports."

"Will I die?" was Kamola's first question.

"No, my dear. You do have something called breast cancer. It is not an easy disease to deal with. But we will do our best. You have to promise that you will do everything I ask you to do."

Now Kamola's father spoke for the first time.

"Please tell me the truth, doctor. Will my girl be cured? What do I need to do? My sons and I will work harder to meet all the expenses for the treatment. But please make her live."

"I will try my best. Kamola, I will include you in a clinical trial with many other cancer patients. Don't worry. There will be no toxic medications or injections. We will give you a special kind of tomatoes every day. You will have to eat one or two of those tomatoes three times a day. It will be tasty and good for you."

"For how long?"

"As long as I say."

"Can I make chutney or curry out of them?"

"You can't cook them. You have to eat them raw. See, you are not the only one who is going to eat them. There are about thirty other patients who will try the same thing."

"Will those tomatoes help me get rid of what I have? And one more question, did my mother have the same thing?"

"To answer your first question, we will see the results soon. To answer your second question, it is very likely that what you have in your breast, your mother had it in her abdomen."

16

SELENA

Sunflower House

"It sounds so different here than at the hospital," said Teresa.

She sat across the meeting table from Selena and Janine. All three women turned their heads toward the open windows of Selena's office where tall sunflowers swayed in the hot September breeze. They heard children playing on the swings and slide and running through the sprinkler on an early fall day that still felt like summer. Janine's daughter Imani raced through the sprinkler, leading Teresa's grandchildren Natan, Christina, Mariana and Gloria. They laughed and jumped in the swaying fan of water and rainbows. Five other boys and girls joined them and three teenagers sat in deck chairs talking and watching over the little

Three Daughters, Three Journeys: Quest for Cancer Cure
Ananda M. Chakrabarty, Jill Charles, Indrani Mondal,
and Ranjita Chattopadhyay
Copyright © 2017 Pan Stanford Publishing Pte. Ltd.
ISBN 978-981-4745-90-1 (Hardcover), 978-1-315-19666-4 (eBook)
www.panstanford.com

ones. Selena knew all these neighbors and their parents by name.

A passerby would think Sunflower House looked like a friendly, welcoming place to live. It looked like many other Hyde Park apartment buildings, with sixty homes behind its four-story walls of red and white Tudor brick. A stranger would never guess that the building housed cancer patients in treatment and their families. In the rooms and halls, women wrapped their bald heads in Polish kerchiefs, African kente cloths, Indian paisley cotton and American cowgirl bandanas.

Lt. Bill Ridge wore his US Marines ballcap and never complained of the pain of chemotherapy, playing poker in the evenings with two other men being treated for colon and prostate cancer. Many supportive husbands, sons and grandsons of the fifty-five women cancer patients stayed here too if their families had lost an apartment or house after paying high medical bills. The only apartment that did not house a cancer patient in recovery was Selena's one-bedroom unit.

Teresa Avila, who had fully recovered from her cancer and gone home to her husband Paolo, now visited several times a week and translated at medical appointments for other Spanish speakers, Mexican and Puerto Rican ladies from Chicago, women who had moved north from Guatemala, El Salvador and Peru for better lives, now fighting for their survival.

Janine taught free art classes to the residents from adults painting watercolors to children making paper bag puppets. Imani came with her and easily befriended Mariana, Christina and the other children.

Gazing at her neighbors, Selena said, "There's no place else I'd rather live."

Selena could have bought a brownstone mansion in Hyde Park, a brand-new Bronzeville townhouse or a 75th-floor condo in the Loop. She could be looking out from a high-rise balcony over Lake Michigan and the skyscrapers, could have bought herself a rose garden in any Chicago neighborhood she liked or any suburb, but that would never be her true home.

Instead of buying a house for herself, Selena used her inheritance and trust fund to renovate a 1930s brick apartment building south of the University of Chicago. Its bankrupt owner sold it cheap because it needed major elevator repair and a new fire alarm and sprinkler system. Selena hired repairmen and learned all she could about lighting, plumbing, painting and carpet, joining in the repairs herself. She moved in and named the building Sunflower House.

At the women's shelter and through Dr. Weiss, Selena found cancer patients too sick to work during their cancer treatment, from elderly widows to young divorced mothers with children who could no longer afford their rent or mortgage payments. Selena rented each Sunflower House apartment for 30 percent of a resident's income; for residents with no income they were free. She qualified for a government tax credit housing program and some of her residents had Section 8 vouchers to help with their rents. Many were homeless when Selena met them. When one patient recovered or died, more were waiting to move into Sunflower House. Survivors and surviving families became friends and helped each other find jobs and share homes when they went back to work and moved on.

Selena remembered what Grandma Grace had told her before she died quietly after ninety-three healthy years. The last time Selena visited Grandma Grace, she told her granddaughter:

"Keep your heart open, Selena. You have your good mind and all your classes to teach and psychology patients. You change so many lives for the better. Never be discouraged and keep on loving the folks you can't save. You won't always be alone, honey. There is someone out there who will take care of you like you take care of everybody else. Keep your heart open."

With all her friends like Teresa and Janine and her dad, cousins, aunts and uncles, Selena never felt lonely. She knew her grandmother had hoped she would marry. Selena still felt attracted to men but had not dated anyone since she and Claude broke up. She threw herself into her work, counseling patients, conducting therapy groups, lecturing at the university and moving residents in and out of Sunflower House. After her dog, Jojo, died, Selena never got another one because she worked such long hours and hated to leave a pet home alone all day.

At the meeting with Teresa and Janine, she sipped her raspberry iced tea and looked at the yellow programs Janine had designed for an art show they were planning. She read:

VITAL ART SHOW
paintings, photos, ceramics, sculptures and textiles
by women and men living with cancer
and in loving memory of artists we have lost

"I can't think of a better way to word that," said Selena. "Not all the artists are survivors."

Three of the paintings in the show were done by Sunflower House residents who had died.

Janine looked down at the color images of painted sunflowers, a pencil sketch of the gray city skyline framed by a hospital window, a photo of 55th Street Beach at sunset, a grandmother's quilt and Col. Ridge's wood-carved bald eagle.

"We'll need to send these to the printer today and to the gallery by Friday," Janine said. "I should have finished these sooner but I've been so busy with the second *Weird Girls* book and Imani's costume for the endangered species play. It's not easy to make a California condor costume out of only black paper and cardboard."

Selena chuckled. "No problem. Imani will make it and I bet Vera can help."

She looked outside for Vera, her twelve-year-old friend, but didn't see her playing outside with Imani and the others.

Teresa examined the program words and said, "How about 'Vital Art Show: paintings, photos, ceramics, sculptures and textiles by artists who have lived with cancer'? That's true of everyone in the show, the survivors and Nadia, Kim and Joan."

Selena nodded and said, "Yes, let's change it. The rest of the design is perfect."

"Paolo can drive our van to the gallery opening," Teresa said. "We can bring eight people."

"I can take five in the minivan," said Janine. "Besides Imani and me."

Selena felt proud of Janine for teaching her daughter that cancer is something friends help each other through, not something too fearful to speak about.

"We have twenty-two people so far who I know will attend and rides for sixteen of them. Liz and Carla will take the El train. I'll ask Martin about a van for the rest. That should take care of everybody."

"You should ride with Martin," Janine said. "To get there early, I mean."

Glancing from Janine to Selena, Teresa tried to suppress a smile.

"Martin always helps out," Selena said softly.

"Well I should round up Natan, Christina, Mariana and Gloria to get them home by dinnertime," Teresa said.

"And I'll go pull Imani out of the sprinkler." Janine added.

She and Teresa hugged Selena goodbye.

Selena put the meeting notes on her desk and took their iced tea glasses to her kitchen sink to wash. She looked out from her apartment windows but still didn't see Vera around. Walking down the hall of Sunflower House, Selena thought about Vera and her dad, Martin.

Martin was a good friend now, the man she trusted most after Daddy. Selena had met Martin at the University of Chicago. One afternoon she had to catch a flight from Midway airport to San Francisco right after her class, to give a presentation about Sunflower House at a cancer survivors' conference. As Selena rushed out of a gray stone building, dragging her rolling suitcase across the university lawn, she spotted a blue and silver taxi and flagged it down.

The driver parked as Selena dashed across the lawn, and got out to help her with suitcase, calmly asking "Could you use a cab?"

"Yes, please," Selena said, catching her breath. "I've got to get to Midway before five."

He was a Puerto Rican man with a big smile and curly black hair.

"Hop in," he said. "I'm your man."

Swiftly yet safely, he drove to the freeway and south toward Midway airport, chatting with Selena on the way.

"I'm sorry to be in such a rush," she explained. "I have a flight at six-forty-five for a cancer survivors' conference in San Francisco."

"Don't be sorry," the driver said. "That's a good cause and I'll get you there. Are you a doctor or a survivor?"

"Actually neither. I lost my mom to breast cancer and I manage an apartment building called Sunflower House for cancer patients and their families."

"I'm sorry to hear about your mother, but I bet she'd be proud."

"Thank you."

At a red light, he gave her his card and introduced himself "I'm Martin Melendez. If you or your clients ever need a ride, I'll give you a discount. I'm an independent driver."

"Thank you, Martin," she said. "I'm Selena Ramos."

She found a Sunflower House card in her pocket and gave it to him.

"Are you from Chicago?" Martin asked.

"Yes, Hyde Park born and raised. And you?"

"I grew up in Humboldt Park," he said. "But I live in Logan Square now with my daughter Vera. She's ten and goes to a public school near there, one for gifted students. Vera and I are alone; her mom and I split up and she's in jail for buying painkillers illegally."

"I'm so sorry," said Selena. "I hope she can get better and start her life over."

"Me too," said Martin. "Zita loves Vera and I take her to visit her mom every week. She calls her almost every day too, to encourage her. Zita and I were just out of high school and already breaking up when we had Vera, but we always put our daughter first."

In only fifteen minutes, Martin reached Midway airport. Selena paid with a generous tip and shook Martin's hand.

"Thank you," she said. "I'll call you."

Selena would call him several times a week for the next two years. Martin helped Sunflower House residents go to and from chemo treatments, oncologist appointments and even move in and out of Sunflower House. Selena relied on him and could talk to Martin about anything.

He brought his daughter Vera to meet Selena. At ten, Vera was shy with strangers, yet brilliant and completely sure of herself with other children. She began to stay at Sunflower House with Selena while her dad drove residents on errands. Selena watched over Vera and helped her with homework, introduced her to Imani and children living at Sunflower House with their mothers. She made friends easily, helped the younger children with spelling, taught the little girls jump rope rhymes and folded paper airplanes with the boys. Vera had curly black hair and big brown eyes like Martin, the same caramel-colored skin, but she was leaner like her mother and wore glasses with blue wire frames.

As Selena walked back down the hall she stopped in the lounge by the main doors, she found Vera at the coffee table with open history books and notecards fanned out in front of her.

"Hi Vera," said Selena. "You can study outside where the light is better if you want."

Vera smiled and sat back.

"Thanks, Selena, but I want to finish my report before I run through the sprinkler. Do you want to hear my speech about Harriet Tubman?"

"I would love it," Selena said. "She's one of my heroes."

"Like your mom?" Vera asked.

"Yes. Mama loved me so much I always feel like she's with me, giving me courage."

It still hurt Selena to talk about her mother but she had told Vera the truth about how her mother died. Vera understood and empathized more than most adults could. She missed her mom, Zita, a lot but never gave up on her. Each week Vera gave her mom a letter and drawings. Selena had seen the letters: *"I know you can quit drugs. I am proud of you for quitting. Here is the polar bear Dad showed me at Lincoln Park Zoo."*

"I'll come see you in your office as soon as I'm done," said Vera. "Two more minutes."

"OK. You can change into your swimsuit in my bathroom after that."

Selena had a one bedroom apartment in Sunflower House, its only resident who had not had cancer. Yet Selena fought the cancer that took her mother from her, every day, in her work and in her friendships.

As she waited for Vera and typed up appointment notes from clients she counseled, Selena thought about her father's August wedding to Dahlia Jones. Dahlia sang jazz and blues and played piano at the Blue Jay nightclub. She had been a friend of the family for years and Selena always liked her, yet Daddy hesitated to tell Selena about their relationship until they were engaged. Selena had given her father a big hug and said, "I'm so glad you're not alone, Daddy. Mama would be glad you're not alone."

Daddy and Dahlia married in the Blue Jay, an evening wedding with a Baptist minister presiding and a nine-piece jazz band playing at the reception. Dahlia sang "At Last, My Love Has Come Along," then joined Daddy on the dance floor.

Selena had been sitting with her cousin Nicolette Wells, a University of Chicago sophomore. Nicolette was the last of Selena's seven other cousins and aunts to be tested for the BRCA1 and BRCA2 genetic mutations. Four of the others had tested positive for at least one of the mutations. None of them had chosen preventative mastectomy and only one, Marta, had chosen an oovectomy "because I'm done having kids, and want to be here for my sons as long as I can." The younger female relatives feared early menopause, losing their husbands or boyfriends, and not being able to give birth to children.

"My test results came back," Nicolette said. "I have the BRCA1 and BRCA2 mutations like you have."

Selena reached across the table and took her hand.

"I'm sorry, Nicolette."

Nicolette shook her hand off.

"I'll be OK. I'm not going to get cancer and lose my job and my husband like Mom did. I'm going to have the mastectomy and oovectomy like you."

"Are you sure?" Selena asked.

"Positive," said Nicolette. "I watched you have your surgeries with no regrets. You don't even look any different. You told me you don't feel different. I'm not scared of menopause at all; it would be great not to have my period anymore. And I'm sure I don't want kids."

Selena looked at Nicolette, slim and confident in her glittering silver gown, her bobbed black hair and skin shining like deep brown velvet. So smart, so happy, so healthy—now—but Nicolette had changed colleges twice and majors three times from Spanish to biology to marketing. Nicolette switched interests and boyfriends quickly, moved out of apartments after 6 months or a year "just for a change." What if she made an irreversible decision against fertility, and later regretted it?

"Menopause is different for every woman," Selena said. "I expected to have hot flashes, trouble sleeping, even confusion or mood swings for a while. I felt lucky that I didn't. I take a supplement to protect my bones and I've always exercised a lot which helps me sleep and stay healthy."

Nicolette leaned over and said, "Plus, you can still date, right? You have Martin."

They looked across the dance floor, where Martin had just arrived, late after a cab-driving shift, but dressed in a black suit and bow tie.

"Martin's a friend," Selena insisted.

"Whatever you say, cuz," Nicolette said, rolling her eyes.

"We need to talk more about your surgery," said Selena. "We could have lunch with Dr. Weiss and talk about all your options. I know you just got the test results and want to protect yourself now but if there is even a slight chance you would ever want to have a child, you should hold off on the oovectomy. Surgery hurts like hell, takes months of visits to the hospital and if you regret it, there's no going back."

"Do you have any regrets?" Nicolette asked.

Martin approached the table and Selena stood up, to clink their champagne flutes together in a toast to the bride and groom. Martin greeted Nicolette, then asked Selena to dance with him.

They glided slowly together in the blue lights and shadows of the dance floor, Selena's long amethyst dress waving around her. Saxophone and piano notes and other smiling couples surrounded them. It was Selena and Martin's first dance but they moved comfortably together, holding hands. She laid her head on his broad shoulder and felt his heartbeat against hers.

"I'm sorry I couldn't be here for the ceremony." Martin said. "Is it strange to see your dad get married?"

"No. I'm happy for him," Selena said. "I love weddings: the food, the cake, the music, all the relatives and friends coming together to celebrate. The only part I don't like is when the bride throws the bouquet and the single women fight over it to see who will get married next."

Martin chuckled and said, "You could get married any time you wanted, Selena."

Selena lifted her head and looked Martin in the eye. Neither spoke, but both understood what neither had dared to mention until now. When the song stopped, they left the dance floor to get some air on a small balcony on the Blue Jay's second floor. The lights of the Chicago Loop skyscrapers glittered all around them: gold, white and blue window lights, red and green traffic lights in the busy streets and the distant full moon leaving white ribbons in the black waves of Lake Michigan. So much possibility shone before them.

"I don't date anybody because I'm too busy," Selena protested.

"I don't date anybody because I'm waiting for you," Martin said.

Selena held both his hands and they leaned close together and kissed. She felt joyful, loved, beautiful, vibrantly alive and had no idea what to do next.

Now, a month later, in her Sunflower House office, Selena shook off the memory and finished typing her notes. Vera bounced into her office and proudly read her report on Harriet Tubman escaping slavery in the 1800s and returning to the South to lead 300 other people to freedom in the northern states. Selena clapped for Vera, then heard more applause outside her office door.

Martin opened it and stepped in.

Vera rushed to him and hugged him.

"Dad! Do you like my report? Did you hear all of it?"

"Yes, Vera. I stood in the hall and listened to the whole thing. I didn't want to interrupt. I'm so proud of you."

"Dad, can I go play in the sprinkler with Imani? Her mom said she'd wait for me to finish my report and then they'll go get dinner. Selena said I can change in her bathroom."

Selena glanced out the window at Imani and her friends still racing through the sprinkler and Janine waiting somewhat impatiently under a maple tree, probably texting her husband Eli that they would be late.

"Go on outside and have fun," said Martin. "School's back in but it's still hot enough for the sprinkler."

Vera rushed outside.

"She's the best," said Selena. "You're lucky to have such a great daughter."

"You're a good friend to Vera," said Martin.

He sat in the chair facing Selena's desk and reached out to hold her hand.

"I wanted to ask you about something," Martin said.

Now Selena felt nervous. Was it about her and him? "Sure."

Martin took a flyer from his pocket and handed it to Selena. It read:

Call for Presentations
World Cancer Treatment and Prevention Conference
Join 200 oncologists, medical researchers, policy makers,
therapists and survivors
from around the world in Wellington, New Zealand
January 25–30
Learn about the latest cancer treatments, research
and prevention
Now accepting proposals for presentations

"You could present there," said Martin. "Think of the difference you could make."

Selena squeezed his hand, holding the flyer with the other. She had presented about Sunflower House to Chicago housing agencies, San Francisco psychologists, New York housing experts and a Congressional committee in Washington, DC.

"This is the biggest cancer conference in the world," Selena said. "Dr. Weiss told me about it; she wanted to submit a presentation about my surgery and have me accompany her as a case study. It's a great opportunity but I'm not an oncologist or researcher. There must be thousands of presentations submitted."

"The deadline is October 1. You have time. Talk to Dr. Weiss. Maybe you could present together about your surgery and your experience at Sunflower House. She's one of the top surgeons in the US and this apartment home is special—housing just for cancer patients. You have residents here of all races and a lot of different ages, religions and backgrounds, who all cooperate and give each other strength. Plus, you could learn about new therapies that could help everyone here. Do it for them and for you, Selena."

Selena had read about new therapies, alternatives to chemotherapy that might remove cancer cells without harming healthy cells. She thought of Josie, too weak to pick up her children after chemotherapy, of Teresa hiding her pain under a black wig, of Lt. Bill Ridge who gave his health for his country. Could there be a safer treatment? Could medical costs be lowered so cancer patients never had to lose their homes?

"I'll ask Dr. Weiss if we could present together," said Selena. "I'll try."

Martin leaned forward and kissed her.

"You always do," he said, smiling.

17

KAMOLA

The Sunshine

KAMOLA STARTED TAKING THOSE drug tomatoes. Dr. Mishra referred to them as Neel tomatoes. She had an MRI each month. Kamola responded to her treatment really well. In four weeks, the size of her tumor was reduced by half. After several months of consuming the Neel tomatoes, she was free of the tumor. The day her MRI result showed no tumor, she went by the river again. She remembered the spring evening three months ago. The season changed into summer now. A lot had happened in her life since then. Even though it was late morning it was quite dark. Black monsoon clouds hung low in the sky. The treetops looked like a greenish black mountain against the backdrop of cloudy sky. Usually

Three Daughters, Three Journeys: Quest for Cancer Cure
Ananda M. Chakrabarty, Jill Charles, Indrani Mondal,
and Ranjita Chattopadhyay
Copyright © 2017 Pan Stanford Publishing Pte. Ltd.
ISBN 978-981-4745-90-1 (Hardcover), 978-1-315-19666-4 (eBook)
www.panstanford.com

she did not have time to sit by the river in the morning. But it was a Sunday—the only day Shibu, Ramen and their father did not have to work in the field.

Two months ago when she left Dr. Bose's office with her father she had mixed feelings. She did not quite fully realize the nature of her disease. But she felt its serpent-like nature, creeping slowly over her, eventually trying to swallow her completely just like it swallowed her mother and grandmother. No one told her this. But somehow she knew. She realized that if she was left untreated she would die too. But now she would survive. She would survive because of Dr. Rini Mishra and the Neel tomatoes. She knew that too. At that moment she was also happy that she decided to seek medical treatment for her breast pain.

She kept eating those special tomatoes without fail for a year. There were moments when she would be scared. She was uncertain of her future. She wished she studied further so she was able to understand the terminology of her own medical reports. For some unknown reason Kamola was never close to girls her age. But in those moments when she felt sad and worried Samir stood by her. Kamola found a trustworthy friend in him. He told her stories of people who survived fatal diseases. That gave Kamola a lot of mental strength and courage. If he could make some time while he came to visit his family in Palampur he read to Kamola. He read whatever appealed to him, poetry, stories, biographies, travelogues etc. Sometimes Kamola could not make out what he read. But she listened attentively anyway. In her mind she could picture an entire world outside the village where she was born and raised. It was a vague picture. But if it was not for Samir that world would have stayed locked forever. Now she could imagine people with different lifestyles, their struggles and triumphs, their happiness and sorrows. It inspired her to lead a life of her own. She started thinking more about who she was and what she would do once she got better. And that is when she decided she would go back to school and study.

Finally the day arrived. That morning Dr. Mishra told her "Guess what Kamola? Your report tells me that you don't need to worry about eating those raw tomatoes anymore."

"What?"

"Yes. You are perfectly fine for now."

"Does that mean I don't need to lie under the cold machine every month anymore?"

"No. Your last MRI report came out clear. You feel for yourself if the lump is still there where you found it in your breast."

Very gently Dr. Mishra guided Kamola's hand to do the self-exam. Kamola was not sure at first. But she did it. She smiled as she could not feel the knot in her breast any more.

"I have a question for you, Kamola. Will you be willing to come with me to this medical conference where I am going to tell other people your story? It will help so many patients. Like you they will also be free of a terrible disease called cancer. It is so exciting! It will open a new horizon in the history of cancer treatment."

Kamola could not grasp the meaning of the last part of Dr. Mishra's speech. But she realized that she was asking Kamola to go somewhere, leaving Palampur for the first time in her life. She told her that she would think about it and walked out of the hospital building.

She came by the river "Ashru" directly. She pondered over Dr. Mishra's request.

Suddenly she felt a gentle touch at her back. Without looking, she could tell who it was. She shivered a little. In all their time together Samir never touched her like that. Kamola turned around.

"Did you already figure out that it was me?"

"Yes"

"How?"

Kamola did not even bother to answer that question. Some things were best unanswered.

"All right. What have you been thinking?"

"Doctor didi has asked me to come to this big city with her for a few days. I am well now. I don't need to eat those tomatoes any more. She wants to talk to other doctors and patients about me."

"Wonderful, Kamola! But wait a minute. Does that mean you will be gone for a few days?"

"I have been thinking about it. I have not decided yet. It is a big decision for me. As you know, I never stepped outside this village. I don't know anyone who is not from this village."

"How about Shibu, Ramen and your father? What are they saying about it?"

"You are the first one I spoke to about it. I think they will be fine if I want to go."

"But do you? Will you not miss them?" Samir asked. "They will miss you too."

"They will manage without me for a couple of days."

"What if you are gone when I am home?"

"What about it?" she asked.

"I will miss you, Kamola. I can't imagine Palampur without you anymore."

"Really?"

Kamola lifted her eyes to Samir's. She saw a swirl of emotion there. Without thinking, she wrapped her arms around him. Immediately he brought her close to his chest. They held each other tight for a few minutes. There was something positive and promising in that embrace. The cloud started dispersing. Through it the golden rays of the sun came out bright and strong. Together Kamola and Samir looked at the river. It was a magical moment. For a brief instant "Ashru," an unknown village river, seemed like a river of liquid gold.

18

The Quest Begins

THE WORLD CANCER CONFERENCE was in Wellington, New Zealand this year and Eryk was looking forward to it. Ever since he had moved to Central America he had been really busy with teaching and his hardcore research job to do more than scan the papers presented at this annual conference just to keep track of the latest developments.

He always carried with him the memory of Dr. Sam Roy's talk at one such conference many years back that had inspired him in his present research endeavor. He looked back at that after-conference dinner with the genial, down-to-earth scientist who had travelled far and wide, literally and figuratively in his efforts to find cancer cure. As everyone was raising a toast to his unforgettable legacy of a cancer cure, Dr. Roy had

Three Daughters, Three Journeys: Quest for Cancer Cure
Ananda M. Chakrabarty, Jill Charles, Indrani Mondal,
and Ranjita Chattopadhyay
Copyright © 2017 Pan Stanford Publishing Pte. Ltd.
ISBN 978-981-4745-90-1 (Hardcover), 978-1-315-19666-4 (eBook)
www.panstanford.com

candidly admitted the truth of the saying that genius was ten percent merit and ninety percent hard work. He had gone on to tell them how ardently he had slogged over the years and how many odds he had battled till he had succeeded in establishing cancer curative and preventive properties in the protein that he had branded as Neelazin. He had travelled to the remote but now famous Blue Frost Land Research Hospital to finally extract this protein from marine bacteria there. He had mentioned a bestseller book *Bugging Cancer* that recounted his struggles and achievements.

At that dinner Roy had struck up a nostalgic conversation with Eryk's friend and fellow oncology resident Dr. Rini Mishra who was also there. Rini was born in the same part of the world as Roy himself and like him had travelled far in search of a dream, curing cancer. Eryk and Rini were both oncologists from nearby medical schools completing their residency at Midwestern Clinic. In the course of their rounds Eryk had been impressed, amused and awed in different ways by this petite Asian doctor who was motivated, smart and quite a tireless go-getter in her profession. As a person, Rini seemed kind and caring with her patients without being over-emotional as he had noticed some women doctors to be. She had a quaint sense of humor, was very fun-loving in her own way and had a way with words that Eryk had come to adore. Eryk couldn't hide a smile as he remembered the many late nights or early mornings after their hospital shifts when Rini came up with some steaming spicy Indian curry just "to wake up the Nordic giant", as she playfully called him. Both had decided, though not in so many words, to put their relationship on hold for they believed that they had a long way to go in their future research, both were fighting for the same cause, a viable and lasting cancer cure. However they were also aware that their approaches were different. They would talk about this for hours, brainstorming and analyzing, often trying to figure out what research course they should best opt for which would suit their purposes. Beneath their spoken words their unspoken bond of comradery, understanding and mutual care had blossomed naturally and imperceptibly. Eryk had felt many a time that it wouldn't have been hard for him to cross over to something more than friendship. And from Rini's eyes and body language he had been almost positive that the same thoughts had crossed her mind as well. But for all her communicative ways, Eryk sensed in Rini a deep commitment to her conservative heritage values, career commitments and mission in life. He had started loving those precious qualities in her for they set her apart from all the other girls he had met. He would listen spellbound when Rini told him how she had grown up in a Third World country where much needed to be done to bring affordable healthcare to the critically ill, especially in rural areas. That was the objective of her coming to study in the USA so she could go back and share her expertise to help the underprivileged. Eryk in turn would tell her his not so spectacular, or so he thought, background of growing up in Gdansk and getting a scholarship to come study medicine abroad for it had endless career opportunities.

On the way to attending this conference, flashbacks of that milestone after-conference dinner where Eryk had last seen Roy and of course Rini, repeatedly came to his mind. It had been during the final year of Eryk's and Rini's residency. Roy's talk about Neelazin had fired them both up and Eryk and Rini couldn't wait to take this natural miracle remedy discovered by this world renowned scientist a few steps forward.

As a resident doctor at Midwestern Clinic, Eryk had seen how distressed and irritable chronic diabetic patients used to be for they needed to take daily insulin shots. Many of them were in constant pain and prayed for an oral remedy. That is why when he heard Roy's Neelazin was being administered as intravenous injections, Eryk was deeply disturbed and vowed to research and find an oral version of Neelazin.

Rini was born in India, with a deep-rooted sympathy for Third World countries where the economy was not very robust and resources were few. She had wanted Neelazin to easily reach patients from all strata of society, possibly through bioagriculture where mass production would be possible. Their conversation had gone far into the night and Dr. Roy, though somewhat skeptical about their ideas and plans, had promised that they could communicate with him when necessary and get all the important bacterial strains and Neelazin isolation procedures as they needed.

A few months later Eryk had been appointed Senior Professor and Research Scientist in the Department of Medicine at Sanchez Institute and had moved to Central America. Rini had gone to India to join a medical school as a professor trying to link up with NGOs conducting an outreach health fair for benefitting rural areas in India. They had never been able to say any formal goodbyes to each other. Occasionally they'd keep in touch by email but Rini was often in areas where electronic access was not the best. The rapid river of time had flown on and they had just floated apart. But Eryk had seen that Rini, Dr. Rini Mishra, would be presenting a paper also on edible Neelazin as a potent anti-cancer agent at this conference. "Quite a lady!" he thought to himself and felt happy for and proud of this spectacular young fighter. He couldn't wait to congratulate her at the conference and perhaps also pick up where they had left off so suddenly so many years ago. He sincerely hoped that Rini had seen that he, Eryk, would be presenting his paper also and he couldn't wait to share with her the agony at first and ecstasy finally, of coming up with his oral Neelazin especially after its grand success in clinical trials in Poland, particularly with Marzena and many such breast and ovarian cancer patients.

Although Eryk had collaborated a few times with Dr. Roy in the course of his research, he was actually attending this World Cancer Conference after almost a decade. He wished Dr. Sam Roy was there but he was travelling the continents to popularize Neelazin.

The day Eryk was going to deliver his paper dawned cool and calm. Eryk finished his usual breakfast of granola and fruit quickly. Taking a large black coffee, he walked towards the hotel foyer on the way to the conference premises next door. At the revolving door hotel exit, Eryk paused for a second to take in the beautiful morning outside. This downtown area of Wellington was renowned for the very many quaint hotels and eateries alongside a picturesque bay on one side and undulating hills on the other. As he stopped suddenly, someone bumped into him from behind and the hot coffee he was holding spilled on his hand making him swear under his breath and yelp loudly in pain.

"Sorry" said the someone in a rather familiar accent and an irate Eryk turned around to find himself staring at a slim young woman with black-framed glasses and a jade-colored scarf twisted around her thick dark hair to keep them away from her face.

"Rini!" cried Eryk and this time he couldn't care less when more coffee spilled on his hand as he threw his arms around his petite friend.

After the initial surprise and exchange of pleasantries, Rini laughed softly and muttered. "I should have known! The unruly giant still bumps into anything and everything!"

"I'm so sorry" began Eryk, thrilled to find that she hadn't forgotten her old endearment. He whistled softly. "It's pretty amazing you came up with edible Neelazin just like you always wanted, and amazingly as part of a food!"

"You haven't done badly yourself, Eryk. Oral Neelazin as a tablet? You always were looking for oral cures," Rini nodded admiringly.

Eryk shrugged, enjoying Rini's appreciation. It seemed to him that after a long time he could actually tell someone who would understand firsthand, the uncertainty and stress of following an unknown path in an unfamiliar territory but being fortunate enough at the end to find the destination. "Don't you think our papers are going to set off some serious fireworks today?"

And Eryk was right. Eryk's presentation on Oral Neelazin tablets and Rini's presentation on Neelazin-containing tomatoes, along with successful large scale case studies in cancer patients, rocked the conference. Eryk and Rini stood hand in hand and looked at each other. Finally it was time for them to be together. And the cameras rolled and news flashed. "There is much hope today," the headlines announced. "As the world gets ready for a real revolution, the ultimate future cure and prevention of 'the King of Mortal Diseases', as cancer had once been known."

19

The Conference

Kamola felt nervous but exhilarated when the plane took off. As the engines roared to life, Kamola took Rini's hand and leaned back in her seat. The jet gathered speed along the runway, then lifted its nose and wheels off the ground.

Kamola was thrilled to fly, to see the Kolkata airport and city streets and buildings recede as they flew, up, up, turning to soar over the Indian Ocean and South Pacific to Wellington, New Zealand.

The jewel-like Pacific Ocean shone aquamarine beneath them as Rini and Kamola spoke about the World Cancer Conference. Kamola could not wait to tell her father, Shibu, Ramen and Samir about the flight, New Zealand, the conference and the people she would meet there. Dr. Mishra would be with her the whole time and they had assured her father she would be safe

Three Daughters, Three Journeys: Quest for Cancer Cure
Ananda M. Chakrabarty, Jill Charles, Indrani Mondal,
and Ranjita Chattopadhyay
Copyright © 2017 Pan Stanford Publishing Pte. Ltd.
ISBN 978-981-4745-90-1 (Hardcover), 978-1-315-19666-4 (eBook)
www.panstanford.com

travelling. Kamola worried about appearing publicly in front of hundreds of doctors and policy makers from around the world, strangers who spoke every language. But Rini was here with her and they now had an opportunity to save many lives, to inform many people like Kamola about the new Neelazin cure.

She touched the azure blue cotton of her scarf and dress, fabric her mother had woven, that Kamola had sewn into a new travelling dress for the conference.

Rini and Kamola looked at a Bengali program for the World Cancer Conference. Kamola had been studying and could read all the presentation titles and descriptions:

Surgical Cancer Prevention
Oral Neelazin: A Pill to Prevent Cancer
Neelazin Tomatoes: A Cancer Remedy in Food

"I used to work with this doctor, Eryk Cyrek," Rini told Kamola. "We were residents together at Midwestern Clinic and both met Dr. Sam Roy, who discovered the bacteria that produce Neelazin."

"His presentation is the day before ours," Kamola noticed. "Do you think the doctors will accept the new Neelzin cures?"

"They'll have a lot of questions about how it works," said Rini. "And it's normal for any new medicine to be proven effective in trials before it can be available to the public, but Dr. Cyrek and I have the studies to show Neelazin works. Thank you for traveling all this way to speak about your experience."

"I'm happy to help," said Kamola. "You saved me and you have become a good friend as well as a trusted doctor."

At the conference, Selena sat beside Dr. Weiss, three other doctors and two patients at a panel on Surgical Cancer Prevention.

Dr. Weiss addressed the audience.

"The main obstacle for surgical cancer prevention is cost and access to health care. Ideally, young women with a family history of breast cancer should be tested for the BRCA1 and BRCA2 genetic mutations between the ages of fifteen and twenty-five. The high cost of the test is a problem for many, especially women without health insurance. In America health insurance will not always cover the cost of the genetic test and the surgical removal of breast tissue and the ovaries. For unemployed or uninsured women and girls, some nonprofit agencies will pay for the genetic test but the patients and families still struggle to raise thousands of dollars for preventative surgeries."

A New Zealander woman in the audience asked "Is it difficult for young women to make the decision about mastectomy and oovectomy at such a young age? Many women do not have their first child until after twenty-five."

"There are many women who would not consider an oovectomy until after having children." Dr. Weiss answered. "It is a personal choice for each individual, but I wish every person could find out, for free, what their individual risk factors are for cancer, including testing for genetic mutations. There are more mutations like those on the BRCA1 and BRCA2 genes still to be discovered."

"Are there health risks associated with early menopause?" asked a young female doctor from Japan. "I'd like to hear from the patients who had mastectomies and oovectomies; did you feel differently afterward or have any regrets?"

"I can speak to that," Selena answered. "I never regretted my preventative surgeries. The mastectomy was painful but I was fortunate to have insurance and to have reconstructive surgery with my own body fat. I never had to worry about a saline implant that might rupture or need to be replaced or silicon which is toxic. The oovectomy was a much bigger change. I used exercise to prevent weight gain and sleeplessness from early menopause. I was lucky not to have hot flashes or mood swings. I felt so relieved that I wasn't at risk for breast or ovarian cancer, at the very least that I had cut my risk by 90 percent. I lost my grandmother to cancer when I was four and also lost my mother to breast cancer when I was eighteen. I had my surgery at twenty-five. It was easier for me than for some women that age because I knew that I didn't want children."

Selena could not help thinking about Vera. She never wanted to give birth or wished for that experience but her life was better now for knowing Vera and all the children at Sunflower House. She wondered about her cousin Nicolette, so sure now that she wanted a mastectomy and oovectomy. Nicolette had changed her mind on colleges, majors, jobs and homes. What if she changed her mind one day about motherhood?

* * *

Marzena could not wait to see Eryk again. He had never visited her in New York and this conference would be their first meeting in years. Eryk had asked Marzena to speak in his presentation on oral Neelazin, as one of the first patients in its large medical trial in Poland.

Marzena's cancer had never recurred, she had bought a small house in New York with a flower garden and still loved her work as an oncology nurse with Dr. Kim Wong. She travelled back to Gdansk for holidays to see her father, stepmother and Uncle Patrick. Her relatives had visited her in New York too, toured the Metropolitan Museum of Art and Times Square and taken the ferry to Ellis Island to see the Statue of Liberty. Marzena had plenty of friends but had never fallen in love with anyone after Eryk.

He still took up all the space in her heart. She told friends she was too busy to date or hadn't met the right man. She also hoped that when she visited Gdansk he might have time for coffee with her or dinner with her and her family. Eryk was friendly but never drew any closer to Marzena. Often he was busy with work or even traveling to conferences when she visited Poland. Yet he never had a girlfriend, so she had reason to hope he might realize one day that they belonged together. *It has to be now, here,* Marzena told herself, *now or never.*

As she settled into her hotel in Wellington, New Zealand, Marzena emailed her dad to tell him she arrived safely, then looked at an email message from Rose O'Brien. A former child leukemia patient of Marezena's, Rose had now grown up and started nursing school herself, staying in touch with Marzena and becoming a good friend, like a younger sister.

Rose had written "I'm so proud of you going to the World Cancer Conference. You know how it feels to treat cancer and to survive it. When you get home, Brendan wants to interview you for *The New World.*"

Rose had emailed a picture of herself and her older brother Brendan, both freckled with auburn hair, smiling at Marzena from the little screen of her phone. Brendan was thirty-one, closer to Marzena's age than Rose's and a successful reporter for the online and print science magazine *The New World*. He prided himself on breaking stories about new energy sources, space exploration and new medical advances. Brendan wore thick glasses but had a sweet face and big smile framed by a red beard. He always made Marzena laugh and learned a few words of Polish to impress her. Brendan had emailed "Get some pictures and notes from your conference panel. *The New World* cannot afford to fly me to New Zealand (sadly) but wants me to feature your story about Neelazin. When you come home, I'd love to interview you over dinner at the restaurant of your choice."

Marzena smiled. Brendan was a true friend and brilliant reporter who obviously adored her, but still, all her thoughts were on Eryk.

After unpacking, Marzena decided to fight jet lag from her flight from New York by exploring downtown Wellington on foot. She enjoyed the views of the sparkling blue Pacific Ocean and lush green hills around Wellington, the sidewalk cafes and lush planters of bright pink and yellow tropical flowers. She imagined walking hand in hand on the beach with Eryk to collect shells or hiking through woods of tall shady trees to mountain lakes.

While imagining what to say when she called Eryk, Marzena was startled to see him right across the street from her, sitting at a wrought iron table at a sidewalk cafe. The sun shone on his blond hair and glasses; he was smiling, laughing. She rushed across the street toward Eryk, barely pausing to look both ways. As she approached, Marzena stayed behind Eryk so he couldn't see her. Then she noticed someone with him.

Across the table from Eryk sat a pretty, petite Indian woman, talking and laughing with him. She wore a royal blue suit with her long black hair in a neat chignon. They didn't touch or hold hands, but the way they looked at each other and their happiness told Marzena the whole story.

Stung and saddened, she ducked inside the coffee shop before Eryk could see her. She stood at the counter with her back to him, ordering a green tea and sipping the hot bitter liquid without adding sugar or even waiting for it to cool. She would need to meet with Eryk later to prepare for their presentation but it hurt too much to see him with another woman, let alone speak to him now and be introduced to her.

As Marzena looked more closely at Eryk's companion, she recognized Dr. Rini Mishra from the conference program. This Indian oncologist was successfully treating many patients with Neelazin-containing tomatoes. In the conference program, Marzena had read Dr. Mishra's biography and noticed that she completed her residency at Midwestern Clinic at the same time as Eryk, and learned about Neelazin from Dr. Sam Roy. Eryk and Rini had known each other for many years before Marzena met him again in Seattle. Here was the real reason Eryk had never gotten engaged or married. While Marzena waited in vain for Eryk, the man she loved was waiting for someone else.

The next day, Marzena calmed herself as much as possible before meeting Eryk for lunch at the hotel restaurant, overlooking the Wellington harbor from a hill. She wanted to hug him, but he shook her hand instead.

"It's so good to see you, Eryk," Marzena said. "It's been so long and I'm so proud to be part of your presentation about Neelazin."

"You've had a distinguished nursing career yourself," Eryk said. "I've read what your patients have written about you online. I want to speak about oral Neelazin as a prevention for cancer as well as a cure, especially for young patients. You could address the experience of your child patients as well as your own treatment with Neelazin."

Quickly their talk turned to their patients and the presentation, away from anything personal. Marzena and Eryk practiced their speeches and gave each other suggestions and questions they were most likely to be asked by the audience. Strangely, Marzena felt no anxiety at all about the presentation now.

"After the conference I have an interview with *The New World* about Neelazin," Marzena told Eryk.

"What a great opportunity," Eryk said. "This conference will introduce Neelazin to the medical and scientific community in Asia, Europe, Africa and the Americas but magazines like *The New World* can help readers without a medical background understand the new cure."

That afternoon Eryk presented his findings on Neelazin as a cancer cure and as a preventative drug to keep young women from ever developing breast or ovarian cancer. Marzena spoke about her patients struggling with chemotherapy and her own rapid recovery from ovarian cancer in the Neelazin drug trial in Poland.

"Did you have any side effects while taking Neelazin?" a Ugandan doctor asked her.

"No," she told him. "I was surprised at how normal I felt, with no weakness or nausea after taking the pill. I could not have tolerated the chemotherapy injections because I have a phobia of needles. I took a pill three times a day until my tumor was gone and then took Neelazin once a week as a preventative drug for five years. I have stayed cancer-free. The oral cure is a safe option for very young patients whose bones and muscles are still growing; Neelazin only targets cancer cells and it consumes them, leaving healthy cells intact. It can offer an alternative to preventative surgery for girls with a family history of cancer, as I had."

Eryk smiled at Marzena, but also at Dr. Rini Mishra, seated in the front row of the audience.

An elderly female doctor from Canada asked Eryk "How can Neelazin cross the intestinal barrier?"

Eryk answered "The chemical modifications in Neelazin allow it to enter through the intestinal epithelial cells. These modifications are designed to break the tight junctions in those cells, unlike other proteins which are digested in the stomach and can barely enter the bloodstream. Once in the bloodstream, entering through the gut, Neelazin will seek and attack only cancer cells and remove cancer anywhere in the body."

Dr. Weiss and Selena sat in the audience, astonished.

"It's amazing." said Dr. Weiss. "If Neelazin were available as a daily pill, it could prevent cancers from BRCA1 and BRCA2 mutations and many other types."

"If they can afford it," said Selena.

An American pharmaceutical company chairman inquired. "How will Neelazin be patented? Is it considered a human invention or a product of nature because the producing bacteria occur naturally in the Arctic Ocean?"

Eryk replied. "Patent laws vary in different countries. In the US, Neelazin could be considered a product of nature and not eligible for a patent. Note, however, that chemical modifications in Neelazin make it different from the natural Neelazin, making it patentable in the US. We have been able to patent and produce Neelazin in India, where it is considered a human invention, and to export and sell it in other countries such as Poland. Dr. Sam Roy pioneered the use of Neelazin, hoping that it could be an affordable remedy with minimal side effects for patients in every country of the world.

Like Dr. Roy, I have made every effort to keep costs low for patients and protect the quality of our medicine. My Indian colleague, Dr. Rini Mishra, has developed an even less expensive method of delivering Neelazin in genetically modified tomatoes. Since Neelazin is a bacterial product and is not known to occur in foods like tomatoes, even Dr. Mishra's tomato Neelazin is patentable in the US."

As Eryk and Rini beamed at each other, Marzena's jealousy simmered. She hoped her face didn't blush and she smiled as the audience clapped, then rose in a standing ovation. Eryk left the conference hall with Rini and this time, Marzena did not follow him.

Back at the hotel, she texted her dad, stepmom, Uncle Pat and then Brendan O'Brien. She wrote to him "I have a lot of pictures and the full transcript of Eryk Cyrek's presentation and my question and answer notes. Maybe we can do our interview over oolong tea and sweet and sour chicken in Chinatown?"

In spite of the night and day time difference, Brendan immediately texted back from New York. "Be sure to get the update on GM tomato Neelazin tomorrow. GM food is controversial but GM Neelazin could be the cheapest cancer cure ever. This is a significant new invention and I can't wait to speak with you in person about it. Best wishes and safe travels, Brendan."

Marzena texted back. "I'll send you all the news tomorrow and look forward to our interview. Best wishes from New Zealand to you and Rose, Marzena."

As much as she dreaded hearing Dr. Rini Mishra speak, Marzena knew Brendan was right. More and more, she looked forward to this interview.

As Kamola and Dr. Rini Mishra took the stage, Dr. Sam Roy introduced them. Rini translated the English to Bengali for Kamola, who had been reading more about cancer and its treatments. Dr. Datta also sat with them to give information about the first Neelazin-containing tomatoes.

Dr. Roy introduced their presentation. "I am pleased to present my distinguished colleague, Dr. Rini Mishra. After completing her medical education at Midwestern Clinic, Dr. Mishra returned to her home country, India, to work in a village clinic in Palampur in West Bengal. Determined to bring medical care to remote areas, and to make an affordable cancer cure, Dr. Mishra collaborated with Dr. Datta of JN University in New Delhi to develop Neelazin-containing tomatoes."

Dr. Mishra spoke. "My interest in oncology began with my initial conversation with Dr. Roy. Like Dr. Eryk Cyrek, I became interested in oncology at Midwestern Clinic. We hoped to find a new cancer treatment that is more affordable than chemotherapy or radiation without toxic side effects to the patients' bones and healthy cells. Dr. Roy discovered that the bacterial Neelazin seeks out and fights cancer cells and only cancer cells in the body. I worked on a method of delivering the medicine orally with no need for injections, surgery or even a pill.

Dr. Datta had worked previously on genetically modified vegetables for their nutritional improvements and better shelf-life. However, GM foods have been controversial in India and other countries. We recognize that many people may not approve of the Neelazin-containing tomato because of the genetic modification. However, we believe over time, as the cancer patients and their relatives all over the world see how important such tomatoes are for fighting cancer, the GM tomato with Neelazin will be accepted even by the skeptics.

One raw tomato three times a day for a year was enough to cure my patient Kamola of breast cancer. Like her mother and grandmother, Kamola has the BRCA1 and BRCA2 genetic mutations for breast and ovarian cancer. She developed a breast tumor when she was only sixteen and came to me for a diagnosis and treatment. Kamola was part of the first large-scale trial of plant-based Neelazin approved by the regulatory agencies and the Institutional Ethics Board. Like my other patients, she has made a full recovery and remained cancer-free to this day."

Hands shot up with questions for Dr. Mishra.

A French biologist asked "Are Neelazin tomatoes considered a GM food or a medicine or both? Would they be illegal in a city, province or country that banned genetically modified plants?"

"Yes, Neelazin tomatoes are genetically modified," said Dr. Mishra. "They are not considered a product of nature as the bacterial Neelazin is. This makes the Neelazin tomatoes eligible for a patent in the US, unlike the bacterial Neelazin. Laws about GM food vary from place to place, but we hope that countries that ban GM food might allow Neelazin tomatoes as a plant-based medicine or nutritional supplement. Many GM foods are designed only to profit the company that produces them, such as herbicide or pesticide-resistant plants sold to farmers to force them to buy new seeds every year for their crops. The average consumers don't necessarily see the benefits of such GM plants, unlike the Neelazin tomato."

Murmurs of agreement rippled through the crowd. Dr. Mishra took a question from Selena.

"I've always been scared of GM food," said Selena. "Some of them even combine plant and animal genes, but these Neelazin tomatoes seem to have no harmful side effects on the patients or the environment. As a cancer survivor and psychologist in America, my main worry about cancer treatment is the high cost of care. I operate an apartment complex in Chicago for cancer patients and their families who have lost their homes because of the high cost of treatment. If Neelazin tomatoes could be patented in the US, what would the cost be for the patients?"

"A few dollars a day," said Dr. Mishra. "Even if they are sold with a GM warning label, the Neelazin tomatoes are non-toxic and can be easily grown in most climates or even raised in greenhouses in cold countries."

An Indian doctor inquired in Bengali and then in English "I have a question for you, Dr. Mishra, and your patient. Does the patient need to continue eating a tomato a day for life to stay healthy?"

"Most patients treated with Neelazin tomatoes ate two or three raw tomatoes a day for six months to two years for a full recovery, depending on their stage of cancer when they are diagnosed," Dr. Mishra explained. "Based on their prognosis, some patients continued to take Neelazin tomatoes once a week for up to five years to prevent a recurrence of tumors. But after that, they would only need screening for cancer on a regular basis and have no higher risk of recurrence than patients treated with chemotherapy or radiation."

Kamola addressed the audience "When I first asked Dr. Mishra for treatment I had no idea what my illness was. I feared for my life because my mother had died when I was twelve of a similar illness. Dr. Mishra told me I had breast cancer and that my mother had died of ovarian cancer. When I began eating the Neelazin tomatoes, they looked and tasted ordinary, ripe and red. But I began to feel healthier in less than a month and my tumor was completely gone in a year. I am very thankful to Dr. Mishra for saving my life and to all of you for sharing this information with patients like me in your home countries."

Dr. Mishra repeated in English what Kamola had said in Bengali and the audience applauded them both.

A Brazilian botanist asked "Could Neelazin protein be found in other food plants? Why did you choose tomatoes?"

Dr. Datta answered. "Dr. Mishra was optimistic that we could keep Neelazin genes expressed to produce functional Neelazin in the raw tomato. Many scientists have genetically modified tomatoes before and I am optimistic that we can also add Neelazin to cucumbers and different types of fruits to provide options for patients who are allergic to tomatoes. It is always essential that Neel tomatoes to be kept raw; cooking will denature the protein and may render it nonfunctional."

Marzena sat beside Eryk as he proudly watched Rini Mishra present. Eryk and Marzena had congratulated each other on the Oral Neelazin presentation the day before. Eryk had been so happy about the discovery, he had not noticed anything troubling Marzena and she decided not to tell him she had loved and lost him.

Watching Kamola speak, Marzena decided she could not hate Dr. Rini Mishra, no matter what had happened with Eryk. Rini was not only a brilliant oncologist who would save thousands of lives, she cared more about her individual patients like Kamola than about fame, money or personal success. If Marzena had met Rini at work in a hospital, they might easily have become good friends. She could not blame Rini for her losses.

Dr. Weiss asked Dr. Mishra "Could Neel tomatoes be used to prevent cancer in women with the BRCA1 and BRCA2 genetic mutations even before the cancer appears?"

Dr. Mishra replied. "We are optimistic that Neel tomatoes and oral Neelazin pills can be used to prevent as well as to cure cancer in patients of any age with a dose as low as once a week."

As the presentation concluded with a standing ovation, Marzena took a picture of Kamola, Dr. Mishra and Dr. Datta on her phone and collected her notes for the interview with Brendan O'Brien. She went into the washroom to be alone and sort out her thoughts.

Marzena looked at her reflection in the mirror. She saw a young girl getting ready to live her life. Nothing more, nothing less. Indeed she was thankful, for she owed that life to her childhood heartthrob, Eryk, but she now realized that there was a connotation to her feelings for him and possibly also his for her that was not really love. It would always be true like the North Star that Eryk had given her back the career and lifestyle she was so passionate about, to go on caring for ailing children and bringing them back to health. This was as much her profession as her personal vindication of the purity of youth and the affirmation of everything innocent and beautiful. That was why it meant so much to her. None of this would have been possible without the blessing of health. Eryk had given it to her when she thought she had lost it. But it was here that her mistake lay. It was not he but the knowledge that he had acquired and imparted, his research in the development of new ways to treat cancer, that had saved her. The two were not the same.

Neither Marzena nor Kamola were mere subjects or case studies for these good-hearted scientists trying to prove their respective approaches to cancer therapy. That is why Eryk had always treated Marzena as an actual person to be held dear and deeply respected, for going on this research journey together with him. His thankfulness at Marzena's trust in him had been apparent in each and every word he spoke to her, every gesture he made and would possibly stay that way for a long time. It was this heartfelt courteous care that Marzena had mistaken for love. From her side too she had possibly misconstrued her own early girlhood attraction for this handsome man and later her dependence on him as a doctor, for something more than simple gratitude and easy friendship.

With this realization, Marzena's intense unhappiness seemed to recede like turbulent tidal waves on a moonlit beach. When the unruly waters were gone, the moonlight shone in all its splendor on the rolling sands, sparkling with unforeseen hope for the future in the horizon. Taking oral Neelazin once every week as a preventive measure remained a symbol of Marzena's lifelong attachment to a journey that in so many ways was larger than life itself.

Just as she began to feel better, Marzena heard a sound of someone else crying, near her. She opened the washroom door and stepped out. A young black woman with long braided hair hurried down the hall, away from the conference, covering her tearful eyes with one hand. Marzena knew she might not be able to help her; they might not even speak the same language. Still, she was determined to try. She followed the lady and did not notice Dr. Rini Mishra and Kamola also following her outside into the sunshine.

20

Three Daughters

AFTER THE PRESENTATION BY the three doctors and patients, Selena rushed outside the lecture hall. She didn't want to hear another question about risks and costs of therapies or hear another story of a lost relative. She ached inside with regret and self-doubt, a type of pain she had not experienced for years.

There had been one other day she felt this way, walking on the beach at 55th Street last year. She saw Claude there, carrying his baby boy in a blue sling on the front of his shirt, holding hands with his wife as they stepped into the surf. In the distance, she made sure he didn't see her, slipping behind an oak tree and waiting until they passed. She no longer missed his love; her life had gone on, better than she ever expected. And yet, at that moment, she could not let Claude see her standing there alone.

Three Daughters, Three Journeys: Quest for Cancer Cure
Ananda M. Chakrabarty, Jill Charles, Indrani Mondal,
and Ranjita Chattopadhyay
Copyright © 2017 Pan Stanford Publishing Pte. Ltd.
ISBN 978-981-4745-90-1 (Hardcover), 978-1-315-19666-4 (eBook)
www.panstanford.com

The afternoon sun shone down on Selena and parrots cackled in the trees. Selena crossed the lawn behind the auditorium, far from the noise of traffic in the street and the flashes of news photographers. Spotting a palm tree in the distance, she sat down under it on a bench, covered her face with her hands and cried.

Was it all for nothing? Selena asked herself. *I could have kept my ovaries. I could have had children. I could have married Claude.*

She shook her head, reconsidering all the reasons not to marry Claude. She loved Martin and Vera and she would never have met them, or Teresa and her friends at Sunflower House without her surgeries and their aftermath. She looked up to see Kamola and Marzena, drawing nearer to her, along with Dr. Rini Mishra, who had addressed the crowd in English and Bengali and also spoken to Marzena and her doctor in English.

Trying to make sense of it all, and not caring who heard her, Selena blurted out "I had a mastectomy and oovectomy to save my life, to live past fifty and now I learn that I could have waited a few years and been cured by a simple protein from a bacterium? Why doesn't everyone know about this?"

Marzena and Kamola looked sympathetic. Even if Kamola did not understand Selena's words, she could guess what she was thinking. The two women sat down quietly on either side of Selena on the bench. She looked from one to the other. Rini stood by Kamola and translated to her what Selena had said and Kamola's replies, allowing the three women to speak to each other.

"I have been so afraid of my own body, afraid that I would inherit cancer like my mother and my grandmother and my aunts," Selena said. "I told my whole family about my surgeries. I thought it was the strong choice. I thought it could save us."

Selena's voice cracked like a windshield and the shattering spread a burning heat over her face and filled her eyes with tears. She had not cried so hard or felt so helpless since her mother's funeral.

She felt a hand on her left shoulder and another on her right. Selena glanced at Kamola and Marzena, took a deep breath, and wiped her eyes with a tissue. She hated to cry in public in front of people she just met.

Kamola's dark brown eyes consoled her and Marzena gave her a small, encouraging smile.

"We know how you feel," Marzena said. "Do not be angry at yourself."

Selena shook her head and spoke to them.

"I feel foolish saying this to either of you. I'm so thankful that you are alive. You took a chance on a new therapy and saved not only your lives, but millions of potential lives. I'm happy for you. I just wish I had a chance to use the new therapy instead of the old one. I wish everyone with cancer had a chance for this."

Rini whispered Selena's words to Kamola, then translated Kamola's words into English as she consoled Selena.

Kamola patted Selena's hand and said, "I know what it is to feel trapped and not have any answers. For years I had no idea what my mother's illness was and when I felt the same symptoms I thought there was no hope for me. You did your best to protect yourself. We are lucky to be alive, all of us."

Marzena nodded.

"I have to tell my cousin Nicolette about the Neelazin," Selena said. "She wanted to have a mastectomy and oovectomy like I had. She has the BRCA1 and BRCA2 genetic mutations, but she's only twenty years old."

"This isn't over yet," Marzena said. "You've helped a lot of people, Selena. You can help your cousin and the women at Sunflower House. We can let cancer patients and doctors around the world know that they have choices. We have all suffered a lot, but I believe it happened for a reason."

She smiled at Kamola and at Rini too. Her body had healed and her heart didn't feel lonely for Eryk anymore. She thought of how proud her mother would be to see her at a world cancer conference in New Zealand, of all the children she could save. She had her whole life to fall in love again, and a chance to have her own children because of doctors like Eryk and Rini and Dr. Sam Roy.

"It will be different for Nicolette," said Selena. "And for both of you. Thank you, Kamola. Thank you, Marzena. Thank you too, Dr. Mishra. I just met you but I feel like we know each other. We have run through the same maze and now we can help others through it."

"We can tell our stories," said Kamola.

"We can save other lives," said Marzena.

"We can make a cancer cure that everyone can afford, in any country of the world," said Dr. Rini Mishra.

The four women joined hands, rose up together and crossed the lawn to meet with the other doctors, scientists and survivors. They had come from different countries and spoke different languages, but now they were allies. Kamola, Marzena, Selena and Rini had all fought cancer in their own ways. Now they would share what they had learned about the new cure, Neelazin, and save lives in every nation. Their story had just begun.

Index

Milton Keynes UK
Ingram Content Group UK Ltd.
UKHW031134141024
449569UK00006B/196